QUALITATIVE INQUIRY AT A CROSSROADS

Qualitative Inquiry at a Crossroads critically reflects on the ever-changing dynamics of qualitative research in the contemporary moment. We live at a crossroads in which the spaces for critical civic discourse are narrowing, in which traditional political ideologies are now questioned: there is no utopian vision on the horizon, only fear and doubt. The moral and ethical foundations of democracy are under assault, global inequality is on the rise, facts are derided as 'fake news'—an uncertain future stands at our door.

Premised on the belief that our troubled times call for a critical inquiry that matters—a discourse committed to a politics of resistance, a politics of possibility—leading international contributors from the United States, United Kingdom, Australia, Spain, Norway, and Denmark present a range of perspectives, challenges, and opportunities for the field. In so doing, they wrestle with questions concerning the intersecting vectors of method, politics, and praxis. More specifically, contributors engage with issues ranging from indigenous and decolonizing methods, arts-based research, and intersectionality to debates over the research marketplace, accountability metrics, and emergent forays into post-qualitative inquiry.

Norman K. Denzin is Distinguished Emeritus Professor of Communications, Sociology, and the Humanities at the University of Illinois, Urbana-Champaign, USA.

Michael D. Giardina is Professor of Media, Politics, and Physical Culture in the Department of Sport Management at Florida State University, USA.

QUALITATIVE INQUIRY AT A CROSSROADS

Political, Performative, and Methodological Reflection

Edited by Norman K. Denzin and Michael D. Giardina

Routledge
Taylor & Francis Group

NEW YORK AND LONDON

First published 2019
by Routledge
52 Vanderbilt Avenue, New York, NY 10017

and by Routledge
2 Park Square, Milton Park, Abingdon, Oxon, OX14 4RN

Routledge is an imprint of the Taylor & Francis Group, an informa business

© 2019 Taylor & Francis

The right of the Norman K. Denzin and Michael D. Giardina to be
identified as the author of the editorial matter, and of the authors for
their individual chapters, has been asserted in accordance with sections 77
and 78 of the Copyright, Designs and Patents Act 1988.

Library of Congress Cataloging-in-Publication Data
A catalog record for this book has been requested

ISBN: 978-0-367-17438-5 (hbk)
ISBN: 978-0-367-17439-2 (pbk)
ISBN: 978-0-429-05679-6 (ebk)

Typeset in Bembo
by Wearset Ltd, Boldon, Tyne and Wear

CONTENTS

ACKNOWLEDGMENTS

We thank Hannah Shakespeare and Matt Bickerton at Routledge for their support of this volume and the larger ICQI project. Thanks are also due Amy Thomas for expert copyediting, Wearset Ltd. for production design, and Lamont Williams for assistance in compiling the index. Many of the chapters in this book were presented as plenary or keynote addresses at the fourteenth International Congress of Qualitative Inquiry, held at the University of Illinois, Urbana-Champaign, in May 2018. We thank the Institute of Communications Research, the College of Media, and the International Institute for Qualitative Inquiry for continued support of the Congress as well as those campus units that contributed time, fund, and/or volunteers to the effort.

The Congress, and by extension this book, would not have materialized without the tireless efforts of Mary Blair, Robin Price, and James Salvo (the glue who continues to hold the whole thing together).

For information on future Congresses, please visit www.icqi.org.

Norman K. Denzin
Michael D. Giardina
October 2018

INTRODUCTION

Qualitative Inquiry at a Crossroads

Norman K. Denzin and Michael D. Giardina

In October of 2018, an article titled "Academic grievance studies and the corruption of scholarship" was published in *Areo*, a small "opinion and analysis digital magazine focused on current affairs." Authored by Helen Pluckrose, James A. Lindsay, and Peter Boghossian (2018), the article recounted the authors' efforts to publish 'fake' papers in leading journals in the humanities and social sciences—especially but not limited to those with a cultural studies, identity studies, or critical theory focus.[1]

In their article, the authors state the intent behind the hoax:

> Something has gone wrong in the university—especially in certain fields within the humanities. Scholarship based less upon finding truth and more upon attending to social grievances has become firmly established, if not fully dominant, within these fields, and their scholars increasingly bully students, administrators, and other departments into adhering to their worldview. This worldview is not scientific, and it is not rigorous. For many, this problem has been growing increasingly obvious, but strong evidence has been lacking. For this reason, the three of us just spent a year working inside the scholarship we see as an intrinsic part of this problem.
>
> *(para. 1)*

In explaining their hoax, the authors detail how they spent a year writing 20 'fake' papers—with falsified (read: made-up) data or empirical material—and the process by which these papers were peer reviewed, accepted, or rejected from a variety of journals in what they pejoratively term "grievance studies," which according to them includes "(feminist) gender studies, masculinities studies, queer studies, sexuality studies, psychoanalysis, critical race theory, critical whiteness theory, fat studies, sociology, and educational philosophy" (para. 17).[2]

The hoax was, not surprisingly, riddled with its own epistemological and axiological issues. Carl T. Bergstrom (2018) astutely captures what the three authors were guilty of from the perspective of academic misconduct:

> Ethically, the project is indefensible. Numerous editors and dozens of unconsenting peer reviewers invested large amounts of time on bad-faith submissions. The hoax, described by its architects as a "reflexive ethnography," appears to lack IRB approval for ethnographic work with human subjects. Two of the four published articles were based on fabricated data or field notes; the fraud was not immediately disclosed. This is straight-up academic misconduct.
>
> *(para. 2)*

Yet this was not an academic study but a deliberate assault on numerous fields of inquiry. More along the lines of the schemes perpetrated by conservative political sensationalists like James O'Keefe than of the scholarly provocation of Alan Sokal (whose 1996 *Social Text* hoax was at least a teachable moment of sorts), this affair raises numerous issues that need to be situated in a broader context.

We should note that we are not concerned with the specific actions taken on the part of various journals that were targeted by the hoax, insofar as what was accepted or rejected, or how the fake manuscripts were handled in the review process. As two people who have been Editors-in-Chief of five scholarly journals over the years,[3] it is not surprising that a weak manuscript made it through the peer review process, or that a particular premise might be regarded as interesting enough to warrant being sent for review rather than desk rejected, or that an external reviewer presented supportive though critical feedback of a poorly developed manuscript. The peer review process is not meant to uncover outright fraud, nor are editors so egotistical as to believe they know everything about every topic. Had the authors of the hoax endeavored to publish fake papers in economics, marketing, medicine, or engineering fields they likely would have had some degree of success, just given the odds (see Drezner, 2018).[4] We *are* concerned, however, with the political context in which the hoax was perpetrated, as well as the conversations that it engendered both inside and outside the academy. For it is illustrative of the current politically charged landscape of higher education, and of critical work being done in in the present moment.

Consider the following: In the aftermath of the hoax's publication, Niall Ferguson (2018) took to Twitter to state: "With their brilliant hoax [the authors] have exposed academic 'grievance studies' for what it is: rubbish masquerading as scholarship." Later, in a *Boston Globe* article titled "Hacking academia," Ferguson referred to the authors of the hoax as "my new heroes" (para. 5) and called their proposition "one of the greatest hoaxes in the history of academia" (para. 6); he also referred to the journals that were targeted as being part of disciplines in which "[t]he rubbish they publish is the counterpart of the rubbish they teach, and the people they teach then graduate with rubbish degrees and live among us" (para. 16). While these comments may read like the ramblings of internet troll, Ferguson is a Senior Fellow at the Hoover Institution at Stanford University and has previously held a named professorship in history at Harvard University. He also served as an advisor to John McCain's 2008 Presidential campaign and has supported other Republican politicians. In other words, he is considered a serious intellectual and conservative thinker who has authored numerous best-selling books on empire, civilization, war, and financial history.

Ferguson was not alone among the conservative intellectual commentariat with his praise for the hoax (and, by extension, its attendant political orientation). Steven Pinker (@sapinker) (2018), the famed Harvard cognitive scientist, tweeted, "Is there any idea so outlandish it won't be published in a Critical/PoMo/Identity/'Theory' journal?"[5] Jordan Peterson, the controversial Canadian psychology professor who has argued against political correctness and claimed that fields such as women's studies, sociology, anthropology, and English literature have been corrupted by neo-Marxism and indoctrination (Levy, 2018), promoted the hoax and offered it up as a necessary corrective to academic leftism. Rod Dreher (2018), a senior editor at *The American Conservative* who has written for the *Wall Street Journal* and *National Review*, referred to the universe of gender studies, feminist theory, critical race theory, and the like as being "bankrupt" and a "sham." Charlie Sykes (2018), a contributing editor at the conservative *Weekly Standard*, exclaimed in a by-the-numbers approval of the hoax that it "exposes the incredible silliness of the academic left." And Ben Shapiro—editor of the conservative website *The Daily Wire* and host of a nationally syndicated radio show—was given space in *Newsweek* to offer his opinion that "A genius academic hoax exposed that liberal arts colleges don't care about the truth," and espouse the belief that "Constructivism is perhaps the most idiotic philosophy at work today in education" (para. 8).[6]

An important summative point to the above reactions can be found at the conclusion of Shapiro's *Newsweek* article. There he writes,

> The authors of this latest hoax have done a real service to those in the general public who still believe that college liberal arts programs search for knowledge rather than reveling in power dynamics. Now the only question is whether parents and students will call for true power to be restored to those who wish to redirect education away from navel-gazing mental masturbation and toward a renewed intellectual search for knowledge.
>
> *(para. 13)*

It is impossible to read that statement as anything other than a full-blown assault on the humanities and some disciplines in the social sciences and, taken to its logical end, as a public pronouncement that these fields *should not exist* (or, at least, not exist in their present form). This hoax and, more importantly, the response to it, is thus, as Natalia Mehlman Petrzela (2018) explains, about "the defunding and devaluing of the humanities—including not just feminist and ethnic studies, but also history, philosophy, literature, and other fields these pranksters would likely deem worthy of continued existence" (para. 7). And though one may scoff at the idea that any of these academic disciplines might become so politically untenable as to be legislated out of existence, that is exactly what recently happened to gender studies programs in Hungary: on October 17, 2018, the government of Hungary effectively banned gender studies degrees, "citing low enrollment numbers that waste taxpayers money and because it is 'an ideology not a science'" (Parke, 2018, para. 1).[7]

At a time when higher education is under constant assault (Giroux, 2007; Slaughter & Rhoades, 2011; Tuchman, 2009), a hoax like this—and its subsequent celebration in some quarters of public discourse—fits into ready-made narratives and is easy pickings for those looking to cast aspersions on the value of a liberal arts education. Indeed, *USA Today*—which has a daily reach of 7 *million readers* and is one of the most highly circulated papers in the United States—allowed the authors to publish an opinion piece (see Lindsay, Boghossian, & Pluckrose, 2018) in which they stated "'Grievance studies' have overtaken academic inquiry into race, gender, and sexuality. Today's scholarship is often just sexism and racism repackaged." Their main argument in this particular advertorial for their hoax is that "we have little reason to trust the concepts coming out of grievance studies"; the concepts specifically named are microaggressions, cultural appropriation, and toxic masculinity, all of which have gained widespread usage in mainstream discourse.[8] Additionally, in the *Wall Street Journal*, which broke news of the hoax in glowing terms, editorial page writer Jillian Kay Melchior (2018) began her article with the statement "The existence of a monthly journal focused on 'feminist geography' is a sign of something gone awry in academia" (para. 1).[9] Even-handed such an article was not.

To be fair, other national news outlets, such as the *New York Times*, *Washington Post*, *Slate*, and *The Atlantic* all covered the issue with varying degrees of criticism and push-back. Daniel W. Drezner (2018), for example, provided astute criticism of the hoax in his *Washington Post* article, allowing that although "[m]ost forms of critical constructivism are not my cup of ontological tea … the belief that much of human society is a social construct is not really that radical a notion" (para. 13), as well as questioning the ethics and methods used to perpetuate the hoax. *The Chronicle of Higher Education* also compiled several pro/con opinion pieces to debate the issue, and academic Twitter was alive with conversations over the issues. It would seem that the authors of the hoax suc-ceeded in generating the kind of discussion they seemingly wanted: that of widespread denunciation or at least 'problematization' of such fields of inquiry on a popular scale, especially *outside of academia*. Drezner (2018), in fact, con-cludes that one (political) outcome of the hoax is that the framing of many of these disciplines as 'grievance studies' "will be adopted far and wide in conser-vative quarters" (para. 14) to attack work being done in the humanities and social sciences. (It already has.) Put differently, their fraud perpetrated epistemic violence (see Spivak, 1998) against a wide swath of critical disciplines, while perpetuating existing divisions within and against such scholarship.

At a Crossroads[10]

We began this Introduction with a brief look at the *Areo* hoax because qual-itative researchers clearly have skin in this game. Beyond the topical studies that look at race, class, gender, sexuality, and the like, several fake manuscripts claimed to use ethnography, autoethnography, and narrative/poetic inquiry in the research design (such as those that were submitted to *Qualitative Inquiry* and *Journal of Contemporary Ethnography*). In terms of framing, Pluckrose et al. (2018) specifically discussed how "Questionable qualitative methodologies such as poetic inquiry and autoethnography (sometimes rightly and pejoratively called 'mesearch') were incorporated" (para. 16) into their fraudulent manuscripts—as if method alone was a disqualifier for scientific inquiry.

We are nearly a half-century removed from the methodological conflicts of the 1970s and the 1980s, and still a litany of familiar criticisms circulate both inside and outside academia. A partial list of, shall we say, grievances, levied against this kind of work:

> Qualitative Inquiry is nonscientific.
> Qualitative Inquiry is fiction.
> Qualitative Inquiry is soft journalism.
> Qualitative Inquiry is political.
> Qualitative Inquiry has no truth criteria.
> Qualitative Inquiry is armchair inquiry.
> Qualitative Inquiry is an anything-goes methodology.
> Qualitative Inquiry is romantic postmodernism.
> Qualitative Inquiry only yields moral criticism.
> Qualitative Inquiry only yields low-quality research results.
> Qualitative Inquiry only yields results that are stereotypical.
> Qualitative Inquiry only yields results that are close to common sense.
> Qualitative Inquiry signals the death of empirical science.
> Qualitative Inquiry is an attack on reason and truth.
> Qualitative Inquiry is not rigorous.

Qualitative Inquiry is not systematic.

Qualitative inquiry lacks an objective methodology.

Qualitative Inquiry does not yield causal analyses.

Qualitative Inquiry does not use randomized controlled experiments.

Qualitative Inquiry does not produce work that can be replicated.

Qualitative Inquiry does not produce work that can be generalized.

Qualitative Inquiry has no well-defined variables.

Qualitative Inquiry produces no hard evidence.

A half-century? Yes. Same criticisms? Yes. Any change? Yes. What? In the traditional and golden ages of qualitative inquiry, positivism reigned. All inquiry was judged against a narrow set of criteria—objective, valid, reliable, accounts of the "Other," and his or her way of life.

Today, that picture has been shattered. The myth of the objective observer has been deconstructed. The qualitative researcher is not an objective, politically neutral observer who stands outside and above the study of the social world. *But that doesn't mean he or she is simply making things up!* Rather, the researcher is historically and locally situated within the very processes being studied (as are all researchers, regardless of paradigm subscribed). A gendered, historical self is brought to this process. This self, as a set of shifting identities, has its own history with the situated practices that define and shape the public issues and private troubles being studied.

In the humanities and social sciences today, there is no longer a God's-eye view that guarantees absolute methodological certainty (see, e.g., Silberzahn et al., 2018. All inquiry reflects the standpoint of the inquirer. All observation is theory-laden. There is no possibility of theory- or value-free knowledge. The days of naïve realism and naïve positivism are over. The criteria for evaluating research are now relative. A critical social science seeks its external grounding not in science, in any of its revisionist, postpositivist forms, but rather in a commitment to critical pedagogy and communitarian feminism with hope but no guarantees. It seeks to understand how power and ideology operate through and across systems of discourse, cultural commodities, and cultural texts. It asks how words and texts and their meanings play a pivotal part in cultures' "decisive performances of race, class [and] gender" (Downing, 1987, p. 80).

This is a dangerous proposition for many to accept politically, for as Lauren Berlant (1997) wrote nearly two decades ago:

The backlash against cultural studies is frequently a euphemism for discomfort with work on contemporary culture around race, sexuality, class, and gender. It is sometimes a way of talking about the fear of losing what little standing intellectual work has gained through its studied irrelevance (and superiority) to capitalist culture. It expresses a fear of popular culture and popularized criticism. It expresses a kind of antielitism made in defense of narrow notions of what proper intellectual objects and intellectual postures should be.

(p. 265)

As some of the public response to the *Areo* hoax reveals, both the political and academic criticisms of qualitative and interpretive research remain, despite visible gains made over the last few decades (e.g., international conferences, high-impact journals, handbooks, graduate programs, etc.). Here the politics of evidence, the politics of research, and a slavish adherence to methodolatry collide once again—a privileged hierarchy of inquiry we have been writing against for at least the last 15 years (see, e.g., Denzin & Giardina, 2006, 2008, 2015; Denzin, Lincoln, & Giardina, 2006). This on top of (or within) the often-paralyzing minutiae of an audit culture (i.e., university accountability metrics, funding pressures, and administrative managerialism; see Denzin & Giardina, 2018; Spooner & McNinch, 2018) that calls into question the value and institutional worth of certain forms of inquiry. Such a climate, we have highlighted previously (Denzin & Giardina, 2017), is alive and well in tenure battles, extramural grant funding agencies, Institutional Review Boards, and so forth—a climate in which qualitative researchers have been forced to respond so as to position themselves to better "attain success and advance within the neoliberal university" (p. 11).

Ironically, as the far-right rages against constructivism and postmodernism, it has simultaneously perverted these concepts for its own political ends. Namely, the subversion of truth in the context of the Trump presidency—a subversion that has its recent political antecedents during the George W. Bush presidency of the early 2000s. Some readers may recall the unnamed Presidential advisor's claim in 2004 that "We're an empire now, and when we act, we create our own reality" (quoted in Suskind, 2004, p. 51).[11] Or comedian Stephen Colbert's (2005) popular mocking of "truthiness," or, the notion that emotion and gut feeling were supplanting reason and evidence in presidential decision making. In the mid-2000s, we ourselves engaged in a discussion of scientifically based research and the war on truth; specifically, as applied to Bush administration policy (see Denzin & Giardina, 2006; Denzin, Lincoln, & Giardina, 2006). Speaking to the subversion or erasure of truth under Trump, Carlos Lozada (2018) remarks that "When truth becomes malleable and contestable regardless of evidence, a mere tussle of manufactured narratives, it becomes less about conveying facts than about picking sides, particularly in politics" (para. 13). That is the critique being levied against 'grievance studies' in the *Areo* hoax. But there is a fundamental intellectual dishonesty at play, for it presumes qualitative and interpretive researchers are ignoring the evidence of their own studies (or simply lack evidence at all) and making arguments to the contrary to support some a priori ideological position. *But that is not how critical scholarship works!*[12]

It is increasingly, however, how the political narrative plays out. As Lee McIntyre (2018) chronicles in his book *Post-Truth*, crusaders on the far-right have grasped on to the poststructuralist idea that "science does not have a monopoly on the truth" for their own political ends; twisting reality into whatever fits the moment, lying with impunity, for what is a lie when all truth is but a construction?[13] *It is within this context that we all now live.*

The Chapters

All of the above brings us to the remit of this volume: qualitative inquiry *at a crossroads*. We use the metaphor of a crossroads to highlight that we are at a point in time when crucial questions about the work we do, the way in which we do it, and the conditions under which that work exists need to be revisited. This is especially true considering the changes that are taking place in the qualitative inquiry community itself. When the International Congress of Qualitative Inquiry was founded in 2005 (and the first book in our series was published, see Denzin & Giardina, 2006), the qualitative community was in a different place. As examples, full-scale debate over post-qualitative inquiry was in its infancy; decolonizing methodologies were just beginning to circulate widely following publication of Linda Tuhiwai Smith's (1999) book on the topic a few years earlier; questions over the role of inquiry in the neoliberal university—including as it relates to audit culture—had yet to capture widespread interest; the *SAGE Handbook of Qualitative Research* was only in its second edition (there are now five editions); and the journal *Qualitative Inquiry* had only been in existence for about ten years.[14]

So, it is accurate to say that progress of a sort has been made—that we are in a different space ten or 15 years on. As anyone who has ever attended the International Congress of Qualitative Inquiry can attest, there is no singular, dominant, or over-riding theoretical or methodological perspective, even within disciplines and subdisciplines. It is also accurate to say that we have more work ahead of us, and not just in terms of protecting our epistemological and ontological turf, but of having meaningful, considered, reflective dialogue about the status and consequence of our work and the context(s) in which our work resides. Put differently, it is fair to ask:[15]

- Are we demonstrating in our work and in our action a greater openness to alternative paradigm critiques? To fruitful dialogue between and across paradigms?
- Are we making meaningful headway in engaging with our colleagues in the hard sciences and/or with (post-)positivists in our own discipline, and materially working toward the change we advocate for in our writing?
- Are we too often speaking to the converted within our own areas of inquiry?

- Are we mentoring the next generation of qualitative researchers to think beyond the established, the normative, and explore the endless complexities of doing qualitative research?
- Are we actively seeking change within the university to support and value qualitative research?
- Are we continually asking what our research accomplishes? (see Pelias, 2015).

Qualitative Inquiry at a Crossroads is divided into three parts: Performative (I), Methodological (II), and Political (III). Contributors to each section showcase the diversity of qualitative inquiry in the historical present. A sampling of key ideas circulating throughout the text is illustrative: decentering epistemologies; intersectionality; decolonization; humanism and posthumanism; politics of research; politics of evidence; research in public; bearing witness; emancipation; social justice; agency; critique; conversation. No single volume can ever capture the complexity and diversity inherent in a field as wide-ranging as ours. However, we hope that what follows is useful in attending to the field in all of its multiplicities and iterations—as a snapshot of theories, methods, and inter-pretations of the field today.

Anne Harris and Stacy Holman Jones ("Between Bodies: Queer Grief in the Anthropocene") open the volume by tracing grief—and in particular the notion of queer grief—from the personal and familial to the collective, to the public, the political, and the environmental. In so doing, they argue that not only experiences but also *expressions* of grief are interlinked: they are experiences that require recognition and relationship; an enactment of the inner-outer experi-ence of being alive, an extension of love (as Sara Ahmed would say). To this end, they hope to inspire readers to consider the ways in which grief is a lan-guage and a body, and without interpersonal and collective practice, the system of symbols of grief separate from the emotions and affects of it, and wither as a social practice.

In Chapter 2 ("'I Wanted the World to See': Black Feminist Performance Auto/Ethnography"), Wilson Okello confronts the historic violence laid bare on and through the Black body. Utilizing Joy James's (1999) concept of limbos and weaving examples of Black motherhood in mourning, anti-lynching pam-phlets, and the murder of Trayvon Martin, Okello asks: What is the work of memory? What is the feel of memory? What is the meaning made of (re)hashing memories? And, how do we exist in the bodies we hold in this historical moment. To this end, he presents a limbo performance through his poem "Buried," and the critical resistance that such acts make meaningful.

In Chapter 3 ("Canvasing the Body: A Radical Relationality of Art, Body, and Vibrant Materiality"), Tami Spry orients us to the "accountability of the body," of the interplay between flesh, body, agency, artistic representation, love, and loss. In so doing, she engages with (posthuman) performance studies and its promise for "a radicalizing of relations between semantic-somatic things with indiscreet transient borders and being"—of engaging matter as agentic to realize material belonging. As a performative explication of agency, Spry shows us a way forward toward understanding how posthumanist ontology can refashion or contribute to new iterations of performance.

In Chapter 4 ("Intersectionality in Education Research: Methodology as Critical Inquiry and Praxis"), Venus E. Evans-Winters and Jennifer Esposito bring to a close Part I by exploring how intersectionality might be used as a conceptual framework in critical qualitative inquiry. They ask: Has education research taken up intersectionality in all of its complexities? and, How can the critical framework as methodology and praxis help critical scholars radically excogitate matrixes of domination? Specifically, they consider how power and authority, resistance and confrontation, space and place, and history shape approaches to inquiry and knowledge production.

Lisa A. Mazzei and Alecia Youngblood Jackson open Part II with their chapter, "Voice in the Agentic Assemblage." Positioning voice in a posthuman ontology that is understood as attributable to a complex network of human and nonhuman agents that exceed the traditional understanding of an individual, they explore in their chapter how a posthumanist stance enables a different consideration of the way in which voice is constituted and constituting in educational inquiry. Drawing from the work of Deleuze and Guattari, Barad, and Bennett, they present a research artifact "that illustrates how this posthuman voice is productively bound to an agentic assemblage." Questions raised include: What happens when voice exceeds language and is more than (un)vocalized words emanating from a speaking subject? If the materiality of voice is not limited to sound (i.e., self-present language emitted from a human mouth), how do we account for it? That is, how might the materiality of voice be located in the space of intra-action among human and non-human objects?

In Chapter 6 ("Wondering in the Dark: The Generative Power of Unknowing in the Arts and in Qualitative Inquiry"), Liora Bresler takes up the productive, formative *experience* of unknowing—the interplay between the known and the unknown coming into being. Reflecting on an assemblage of moments from art and music to educational experiences and meditation retreats, Bresler details how various forms of unknowing have proved generative in her scholarly life—proved invaluable for negotiating different roles. This unknowing, she argues, troubles what and how we 'know,' and calls for us to open ourselves up to the processes of becoming that we encounter.

In Chapter 7 ("Virtuous Inquiry, Refusal, and Cynical work"), Aaron M. Kuntz argues for "a collectively engaged and relationally enacted virtuous inquiry built on practices of radical refusal, truth-telling, and an ethical insistence that we might (must) become otherwise." In making this case, he implores us not to become "desensitized to the logics of injustice and inequity" that riddle our historical present—riddle ethical practices and the conditions under which scholarly inquiry takes place. In particular, Kuntz focuses on the Ancient Greek concept of *parrhesia*—the intersection of citizenship, responsibility, and risk—as it relates to truth, inquiry, and its moral conduct.

In Chapter 8 ("Theorizing from the Streets: De/colonizing, Contemplative, and Creative Approaches and Consideration of Quality in Arts-Based Qualitative Research"), Kakali Bhattacharya traces her journey of un/learning colonizing ontoepistemologies in favor of an approach she has termed *theorizing from the streets* to create expansive spaces in inquiry. This journey highlights her departure from both traditional and contemporary understandings of scholarly knowledge-making in qualitative inquiry and identifies what she considers as benchmarks of quality in qualitative work. To this end, she asks: What is considered knowledge making in academia, and who is privileged as a knowledge maker? and, Are we open to recognizing knowledge that exists outside the bounds of academic gatekeeping?

In Chapter 9 ("Stay Human: Can We Be Human after Posthumanism?"), Svend Brinkmann addresses a disconnect between discussion of posthuman philosophy on the one hand, and the materiality of human tragedies on the other hand (e.g., war, famine, poverty, etc.). Reflecting on the growing popularity of posthuman and post-qualitative approaches within the field, Brinkmann wonders how we can "stay human without becoming dangerously anthropocentric." He begins by discussing posthuman philosophy and the related post-qualitative critique of conventional (humanist) qualitative research. He then discusses if there is something 'worth preserving' in humanism, including if it is intellectually feasible now given trends in the field. He concludes by turning to the work of philosopher Hans Jonas, who endeavored to "articulate a kind of humanism that avoids essentialism."

Julianne Cheek opens Part III with her chapter on the political economy of (qualitative) research ("Resisting the Commodified Researcher Self: Interrogating the Data Doubles We Create for Ourselves when Buying and Selling our Research Products in the Research Marketplace"). Reflecting on the 'elephant in the room'—that is, the research marketplace we all inhabit in the neoliberal context—Cheek asks:

> What sort of researcher self am I? How did I come to be that self? Why did I choose to be that self? And, crucially, what self do I want to be, and what might this mean for the way that I think and undertake my research in the research marketplace?

In addressing these questions, she outlines a number of quandaries that (qualitative) researchers face in this marketplace, how they have navigated them, and how we might grapple with—and push back and beyond—these questions about our research and our researcher selves.

In Chapter 11 ("Contesting Accountability Metrics in Troubled Times through Communicative Methodology"), Aitor Gómez González engages with the politics of research and explains how it is possible to simultaneously achieve scientific, policy, and social impacts by applying communicative methodology throughout the entire research process. In so doing, he calls for us to look beyond a dependency on metrics and encourages us to publish not for professional accolades or to advance particular institutional metrics but with an intention for social and policy impact: in other words, publications that demonstrate real social utility, contributing to improving the lives of the people, rather than adding another line to the CV and calling it social justice.

In Chapter 12 ("Seduction and Desire: The Power of Spectacle"), Bronwyn Davies explores some of the implications of the current political situation for qualitative inquiry. To do this, she looks at some of the historical roots of neoliberalism, and of qualitative inquiry, and at the clash between them that is evidenced in the Trilateral Commission's report on 'governability,' which in the 1970s found that citizens had become too radical, thus making democracy *unaffordable*. That is, their protests, the report claimed, were interfering with the flow of global capital. Within that broader historical and political framework, she examines the advent of the Trump presidency, analyzing it as spectacle and as seduction, and explores some possible implications of his presidency for our work as qualitative researchers (whether we are located in the United States or elsewhere). She concludes by charging us to "pry open the dynamics of the spectacle to discover how they work—and how to deconstruct them."

Karen Staller brings the volume to a close with her chapter reflecting on social justice and qualitative inquiry in troubled times ("Stitching Tattered Cloth"). She positions her argument by acknowledging that whereas qualitative inquiry has been used to "expose fault lines and resist oppressions," less often it has been successful in bridging differences, finding common ground, or "stitching seams along frayed edges." Turning to the work of social reformer Charles Loring Brace (1826–1890) and the impact he had on child welfare, Staller chronicles the parallels between 'troubled times' of the end of the nineteenth century and the present day—growing income inequality, unprecedented immigration, religious intolerance, nativist political parties, debates between science and religion, and racial violence—and asks: What can one person possibly do against these odds? Using the example of Brace and the creation of the Children's Aid Society, she asserts that rather than get lost in the chaos of the moment, we "need to see the entirety of [a] project, envision its collective force, and its long-term impact."

By Way of a Conclusion

This is the fifteenth volume with roots in the International Congress of Qualitative Inquiry, and perhaps the most important insofar as asking readers to critically reflect on their own work in relationship the broader questions raised in this introduction. There is a need to unsettle traditional concepts of what counts as research, as evidence, as legitimate inquiry. How can such work become part of the public conversation? Who can speak for whom? How are voices to be represented? Can we forge new models of performance, representation, intervention, and praxis? Can we rethink what we mean by ethical inquiry? Can we train a new generation of engaged scholars and community leaders? What counts as scholarship in the neoliberal public sphere? Can we imagine new models of accountability, how do we talk about impact, change, change for whom?

This is a call for inquiry that addresses inequities in the economy, education, employment, the environment, health, housing, food, and water, inquiry that embraces the global cry for peace and justice. We still have a job to do. Let's get to it.

Notes

1 In the spirit of transparency, we are editors of two of the 20 journals to which fake manuscripts were submitted: *Qualitative Inquiry* (Denzin) and *Sociology of Sport Journal* (Giardina). Neither journal published the fraudulent manuscript that was submitted. Additionally, several of the journals to which fake papers were submitted are published by Taylor & Francis, which is the publisher of this book.

2 As a point of order, it is unsurprising these authors would term these fields of inquiry as 'grievance studies' given their public proclamations: Boghossian has been open about his contempt for gender studies as an academic discipline, which he referred to in a 2016 tweet as "primarily composed of ideologues who view indoctrination as their primary duty. These departments must be defunded"; he later expanded that decree to include any discipline "infected with postmodernism." Similarly, Lindsay has claimed that gender studies, postcolonial studies, and the like "is an assault on the truth, and we've offered overwhelming and undeniable evidence of our claims." And Pluckrose, the editor of *Areo*, stated, "I went deep into feminist epistemology and James [Lindsay] into masculinity studies and Peter [Boghossian] into fat studies. We suffered so you don't have to. Christlike is our sacrifice" (October 10, 2018, Tweet, @hpluckrose). The contempt with which the authors sneer at these fields, calling them (in the case of Lindsay) "the scholarly equivalent of nuclear waste," reveals beyond the pale that the hoax was perpetrated not to engage in considered dialogue with these fields or even raise questions with the peer review process in academia or gauge trends within scholarly publishing, but to assert a privileged sense of 'what counts' (and what does *not* count) as legitimate scholarship in the university.

3 The journals are: *Qualitative Inquiry*; *Cultural Studies ↔ Critical Methodologies*; *International Review for Qualitative Research*; *Sociological Quarterly*; *Sociology of Sport Journal*.

4 In fact, during the writing of this introduction, Gina Kolata (2018) reported in the *New York Times* that a prominent Harvard Medical School cardiologist had published 31 studies with "fabricated or falsified data," all of which Harvard officials hoped would be retracted by the relevant journals that had published the studies.

5 This is not surprising given that he has previously claimed identity politics is "an enemy of reason and Enlightenment values."

6 Lest we forget, *The Daily Wire* has published false, inaccurate, and unverified stories of its own in the past.

7 Although it would seem far-fetched that similar legislative action would be taken in the United States, it is not unheard of for departments in a university to be shuttered due to low enrollment, or for political pressure to be exerted on a university by state

legislatures. When you consider the number of right-wing politicians in the United States who have taken gender studies or ethnic studies to task over the last few years (making both economic and 'scientific' arguments against them), it begins to seem less far-fetched. As well, the US online magazine *The Federalist* (a play on the conservative Federalist Society) essentially endorsed this policy decision, publishing an article by Sumantra Maitra (2018) that wonders aloud if this policy is "worth emulating." As a side note, Hungary's autocratic prime minister, Viktor Orban, signed the decree into law; he has been called the Donald Trump of Europe.

8 Elsewhere they identified white fragility as another concept for which they see no use, and which they imply has negatively impacted society. We disagree.

9 Melchior regularly publishes anti-progressive missives for the *Wall Street Journal* editorial page, including recent pieces that: conflate bell hooks with religious zealots; decry "the left's mania for identity politics"; and mock scholarly writing on popular culture.

10 This section draws from and updates arguments from Denzin, 2017.

11 The unnamed advisor was later revealed to be Karl Rove.

12 As we argued more than a decade ago (see Denzin, Lincoln, & Giardina, 2006), these kinds of arguments

> reproduce a variant of the evidence-based model and its criticisms of poststructural, performative sensibilities. They can be used to provide political support for the methodological marginalization of many of the positions we endorse. In any event, we know few—if any—anti-methodological, Feyerabendian, 'anything goes', romantic postmodernists. Our experience has shown most of our colleagues to be quite rigorous, extremely attentive to method and methodology, and far more hard-nosed than romantic.
>
> *(pp. 774–775)*

13 One might also recall right-wing radio host Rush Limbaugh referring to Antonio Gramsci's concept of hegemony and the capturing of key institutions as a playbook to be followed rather than a warning to be heeded (see Alterman, 2001).

14 Other key moments, Mitch Allen (2015) points out, included the founding of journals such as *International Journal of Qualitative Studies in Education*, *Qualitative Research*, and *Ethnography*, as well as SAGE's Handbook program, which led to publication of more than 75 texts including: *The SAGE Handbook of Interview Research*, *The SAGE Handbook of Action Research*, *The SAGE Handbook of Case-Based Methods*, and so forth. That Routledge and Oxford University Press soon followed with their own methods titles and handbooks speaks to the explosion in demand for qualitative training and research.

15 The following is modified from Denzin, 2017, and informed in part by Guba, 1990, and Pelias, 2015.

References

Allen, M. (2015). Qualitative inquiry and arranged marriage. *Qualitative Inquiry, 21*(7), 645–647.

Alterman, E. (2001). The not obviously insane network. *The Nation*. Retrieved October 11, 2018 from www.thenation.com/article/not-obviously-insane-network/.

Bergstrom, C. T. (2018). What the 'grievance studies' hoax means: A hollow exercise in mean-spirited mockery. *The Chronicle of Higher Education*. Retrieved October 11, 2018 from www.chronicle.com/article/What-the-Grievance/244753.

Berlant, L. G. (1997). *The queen of America goes to Washington City: Essays on sex and citizenship*. Durham, NC: Duke University Press.

Denzin, N. K. & Giardina, M. D. (Eds.) (2006). *Qualitative inquiry and the conservative challenge: Confronting methodological fundamentalism.* Walnut Creek, CA: Left Coast Press.

Denzin, N. K. & Giardina, M. D. (Eds.) (2008). *Qualitative inquiry and the politics of evidence.* Walnut Creek, CA: Left Coast Press.

Denzin, N. K. & Giardina, M. D. (Eds.) (2015). *Qualitative inquiry and the politics of research.* Walnut Creek, CA: Left Coast Press.

Denzin, N. K., & Giardina, M. D. (Eds.) (2018). *Qualitative inquiry in the public sphere.* London: Routledge.

Denzin, N. K., & Giardina, M. D. (Eds.) (2017). *Qualitative inquiry in neoliberal times.* London: Routledge.

Denzin, N. K., Lincoln, Y. S., & Giardina, M. D. (2006). Disciplining qualitative research. *International Journal for Qualitative Studies in Education, 19*(6), 796–782.

Downing, D. B. (1987). Deconstruction's scruples: The politics of enlightened critique. *Diacritics, 17*(1), 66–81.

Dreher, R. (2018). The bankruptcy of grievance studies. *The American Conservative.* Retrieved October 8, 2018 from www.theamericanconservative.com/dreher/the-bankruptcy-of-grievance-studies/.

Drezner, D. W. (2018). A paper that would never have gotten past peer review criticizes the academy. Film at 11. *Washington Post.* Retrieved October 6, 2018 from www.washingtonpost.com/outlook/2018/10/04/paper-that-would-never-have-gotten-past-peer-review-criticizes-academy-film/?utm_term=.241e86e8e67d.

Ferguson, N. (2018). Hacking academia. *The Boston Globe.* Retrieved October 8, 2018 from www.bostonglobe.com/opinion/2018/10/08/hacking-academia/J1kwFraoKXvmycfAJk2wVM/story.html.

Giroux, H. A. (2007). *The university in chains: Confronting the military-industrial-academic complex.* New York: Routledge.

Guba, E. (1990). Carrying on the dialog. In E. Guba (Ed.), *The paradigm dialog* (pp. 368–378). Thousand Oaks, CA: Sage.

Kolata, G. (2018). Harvard calls for retraction of dozens of studies by noted cardiac researcher. *New York Times.* Retrieved October 17, 2018 from www.nytimes.com/2018/10/15/health/piero-anversa-fraud-retractions.html.

Levy, S. (2018). Jordan Peterson: Certain university disciplines 'corrupted.' *Toronto Sun.* Retrieved October 8, 2018 from https://torontosun.com/2017/06/29/jordan-peterson-certain-university-disciplines-corrupted/wcm/9189041e-131b-4fb7-a501-0136e86790f3.

Lindsay, J. A., Boghossian, P., & Pluckrose, H. (2018). From dog rape to white men in chains: We fooled the biased academic left with fake studies. *USA Today.* Retrieved October 11, 2018 from www.usatoday.com/story/opinion/voices/2018/10/10/grievance-studies-academia-fake-feminist-hypatia-mein-kampf-racism-column/1575219002/.

Lozada, C. (2018). Can truth survive this president? An honest investigation. *Washington Post.* Retrieved October 4, 2018 from www.washingtonpost.com/news/book-party/wp/2018/07/13/feature/can-truth-survive-this-president-an-honest-investigation/?utm_term=.3277ee1c4060.

McIntyre, L. (2018). *Post-Truth.* Cambridge, MA: MIT Press.

Maitra, S. (2018). Hungary cuts taxpayer funding to inane gender studies departments. Retrieved from https://thefederalist.com/2018/10/18/hungary-cuts-taxpayer-funding-inane-gender-studies-departments/.

Melchior, J. K. (2018). Fake news comes to academia. *The Wall Street Journal*. Retrieved from www.wsj.com/articles/fake-news-comes-to-academia-1538520950.

Parke, C. (2018). Hungary bans gender studies because it is 'an ideology not science.' *Fox News*. Retrieved October 18, 2018 from www.foxnews.com/world/hungary-bans-gender-studies.

Pelias, R. (2015). A story located in 'shoulds': Toward a productive future for qualitative inquiry. *Qualitative Inquiry, 21*(7), 609–611.

Petrzela, N. M. (2018). What the 'grievance studies' hoax means: A limited intellectual vision. *The Chronicle of Higher Education*. Retrieved October 11, 2018 from www.chronicle.com/article/What-the-Grievance/244753.

Pluckrose, H., Lindsay, J. A., & Boghossian, P. (2018). Academic grievance studies and the corruption of scholarship. *Areo*. Retrieved October 4, 2018 from https://areomagazine.com/2018/10/02/academic-grievance-studies-and-the-corruption-of-scholarship/.

Shapiro, B. (2018). A genius academic hoax exposed that liberal arts colleges don't care about truth. *Newsweek*. Retrieved October 8, 2018 from www.newsweek.com/ben-shapiro-genius-academic-hoax-exposed-liberal-arts-colleges-dont-care-1155013.

Silberzahn, R. et al. (2018). Many analysts, one data set: Making transparent how variations in analytic choices affect results. *Advances in methods and practices in psychological science, 1*(3), 337–356.

Slaughter, S., & Rhoades, G. (2011). *Academic capitalism and the new economy: Markets, state, and higher education*. Baltimore, MD: Johns Hopkins University Press.

Spivak, G. C. (1998). Can the subaltern speak. In C. Nelson and L. Grossberg (Eds.), *Marxism and the interpretation of culture* (pp. 271–313). Urbana, IL: University of Illinois Press.

Spooner, M. & McNinch, J. (2018). *Dissident knowledge in higher education*. Regina, Saskatchewan: University of Regina Press.

Suskind, R. (2004). Faith, certainty, and the presidency of George W. Bush. *New York Times Magazine*. Retrieved October 4, 2018 from www.nytimes.com/2004/10/17/magazine/faith-certainty-and-the-presidency-of-george-w-bush.html.

Sykes, C. (2018). The unbearable silliness of the academic left. *The Weekly Standard*. Retrieved October 10, 2018 from www.thecontrarianconservative.com/blog/2018/10/8/the-unberable-silliness-of-the-academic-left.

Tuchman, G. (2009). *Wannabe U: Inside the corporate university*. Chicago, IL: University of Chicago Press.

PART I
Performative Reflections

1

BETWEEN BODIES

Queer Grief in the Anthropocene

Anne Harris and Stacy Holman Jones

> There is a time in life when you expect the world to be always full of new things. And then comes a day when you realise that is not how it will be at all. You see that life will become a thing made of holes. Absences. Losses. Things that were there and are no longer.
>
> *(Macdonald, 2014)*

Writing Bodies in Relation

The sulphur-crested cockatoos squawk an echoing line through the sky of the Yarra Valley old-growth eucalyptus forest. We are gathered here on a writing retreat, making space to make something out of nothing. Words. Friendship. Community. Poetry. We make space to retreat from the world and its neoliberal demands, the disciplining machine of capitalism and the chaotic order it imposes to tend to emotional and relational landscapes. Writing bodies in relation.

I stare out the window of the 100-year old farmhouse while I write, my fingers dancing along the keyboard. Typing is a tactile and embodied practice for me. From my years of training as a pianist and 'touch typist' from the age of four (thanks, Mom), I can look out at the world while my fingers trace the symbols of meaning that my brain and memory produces. Today I look at the white birds with yellow head crest feathers, as they cluck and sweep through the sky in this valley.

The old gum forest offers infinite greens. On a morning walk, I recalled how the forest in New York is so different, has such a narrower palette of green, compared to the wide-ranging grey- and silver-greens of the Australian bush. The greens I didn't notice while growing up in the woods outside of Albany. The greens that are buried, carried, somewhere deep in my formative cells when I was just learning what green was, when I was learning what it meant to go play in the woods, what it meant to find church in nature, not in a building.

When I think about home, about childhood, about that which is now past, I think mostly about place, and mostly about the outdoors, which I have written about but still don't understand. Memories are tied up with, bounded by, and perhaps constructed in the spaces and places where they occurred, and the objects with which they were enacted. For me, the outdoors and the beautiful natural landscape of the upstate New York woods behind my house are inextricably bound up with my childhood, lost family members, dead animals. Bound up with that sense of wonder and visceral aching loss and discovery sitting together with the comfort of the familiar smells, objects and animals of the woods. It is part of both the joy and also the grief that I have when I recall or revisit my family and my childhood.

Going beyond domestic animal relations, and personal connections with nature, others have written about the links between human-centred grief and collective grief at the devastation of the planet (typifying and theorised as the Anthropocene). Some (like us) also theorise and collectivise about public grief, collective mourning, and an *activist affect* in which we are all caught up (Harris & Holman Jones, 2018). In this essay, we see ourselves as bodies among and in relationship with other bodies (animal-bodies, thing-bodies, plant-bodies) that help us to understand our own individual grief, suffering and mourning, as interconnected with the degradation of the planet, a continuum of alienation that has reached epic proportions.

The Beauty of Grief and Loss

We map the traces of our grief with objects, scenes. We remember the illnesses and deaths by the room, the time of day, the pillow under a mother's hand, the smells, the light streaming through the windows.

My mother, Anna Mae Harris, died on 25 August 2007 in Evansville, Indiana, the city where she was born. She died in her hospice bed, when my brother and I went out for a smoke.

Eleven days earlier, she had one final stroke
after having one final breakfast,
and was rushed to the hospital on a Friday morning.

The family stood around her bed in the emergency room and the consulting doctor pulled back the curtain after reading her slides. He spoke directly to my mother.

'This is a catastrophic stroke,' he said. 'There's no way for us to stop the bleeding in your brain. It's still bleeding now. Anna Mae, I'm afraid you're going to die from this.'

Now I'm a blunt person, but this level of honesty took my own — and our collective — breath away.

I touched my mother's legs. 'Do you understand what the doctor's saying?' I asked.

She nodded. 'I'm going to die', she said, her speech a bit slurred. These would be her last words.

My turn to nod, blinking back tears. 'That's right.'

Her sister Cyrilla stroked her face. 'It's okay honey, it's okay. We're all here together.'

My mother smiled a small child smile, kept nodding weakly.

We all cried except my mother, who sat there blinking from the outside and seemed to have been waiting for this moment all along.

Rebecca Solnit tells us that

> Stories of suffering and destruction are endless and overwhelming these days, and you cannot respond to all of them. If you don't shut them out entirely, you must choose which to respond to and how to respond via both affective and deliberative processes.
>
> *(2013, p. 19)*

When we cut off the pain of grief both individually and culturally, we cut off its beauty and the possibility of solidarity and of sharing in its experience. Like gathering, grief can be poetry; indeed, it often leads us there. It is the cut that breaks the flow of normative life. Solnit again reminds us of the affect of grief, it's visceral power and chaos, and yet its naturalness:

> Weeping like ice melting, like winter snow turning into spring rivers, a spring that comes as grief, as waking up to suffering that is the beginning of doing something about it, weeping tears of affection and loss that are always hot and sometimes make roses grow.
>
> *(2013, p. 190)*

Perhaps as a symptom of an insistently binary culture that parses gender, race, age, ability and research paradigms (among other things) into black and white, either-or structures, we also binarise joy and sorrow. Solnit, Sarah Ahmed, and others remind us that there are many creative and bountiful points of intersection in our experiences of grief — for example the happiness inside sadness, the joys of sorrowful sentiments, and the ebb and flow of emotion. For Solnit,

> Sadness always contains distance, spaciousness, takes us away, while happiness at best brings us home to this very moment, this very place, so perhaps they are the sentiments of the far and the near (though rage and fear arise from the proximity of the unwanted as well as the absence or

departure or threat of departure of the desired). Sadness and happiness – if those are even useful words, because as the years have gone by I have wondered if we want other language for emotion, if we would rather speak of deep and shallow, because the things that move people to tears are sometimes joyous and because the attempts to ward off sadness so often ward off depth instead – by distraction, for example.

(2013, p. 266)

And what sort of distractions do we seek, desperate to ward off sadness, depth and death?

My father had his first heart attack when he was in his forties. I'd just returned to school for my master's. My mother called me in California from North Carolina late on a Sunday night. Her voice was thin, drained of emotion.

'Your father had a heart attack last night', she said. 'He's going to have open heart surgery on Tuesday. I know you're busy with school. I just wanted you to know.'

'I'm coming', I said.

'No', my mother said. 'It's not necessary.'

I went.

On the plane I read Aristotle's *Poetics*, the book assigned for the rhetoric seminar I'd miss on the day my father had surgery. I thought that Aristotle was very certain and very mathematical, just like my father. That certainty and logic calmed the knot that formed in my throat every time I thought of him. Every time I wondered what I was doing in graduate school. Every time I told people I was leaving the surety of my office job to study rhetoric and performance, most importantly when I thought of telling my father.

I read the *Poetics*, counting down the hours and minutes until he came out of surgery. I loved that Aristotle loved poetry and the theatre and wanted to say a thing or two to Plato and anyone else who wanted to banish emotion and drama from its important place in public life as if these things did not count. As if the relationships we make in language and in bodies gathered together to make something out of nothing are not real. Or simply are not. In these respects, my father was *not* like Aristotle.

I read the *Poetics* sitting next to my father in the recovery room, trying to block out the sounds of the monitors connected to every part of his body. I thought Aristotle was an awful heterosexist and looked over at my father, who was of the opinion that women (and certainly queer women), or at least this queer woman, were not made for certain things – mathematics, the useless study of theatre and poetry, graduate school. Still, I wondered if my father might change his opinion of me if I read aloud the hateful things Aristotle wrote about women. I asked him if he wanted me to read my first book of graduate school to him. My father shook his head no.

The Queer Objects of Grief

Not all grief – or love – is equal in the social sphere. Queer grief is a

> a lesser tie, a tie that is not binding, that does not endure in matters of life and death. The power of the distinction between friends and family is legislative, as if only family counts, as if other relationships are not real, or are simply not. When queer grief is not recognized, because queer relationships are not recognized, then you become 'nonrelatives,' you become unrelated, you become not. You are alone in your grief. You are left waiting.
>
> *(Ahmed, 2010, p. 109)*

In *Cultural Politics of Emotion*, Sara Ahmed (2013) discusses public grief, arguing that not all grief, not all public losses, are equal. Rather, the sociality of grief is measured in terms of proximity and whether 'others' 'would want collective grief to be extended to them … when such a grief might "take in" what was not, in the first place, "allowed" near' (footnote 12, p. 176).

Ahmed explores public grief and other emotions as a shared object, drawing on Max Scheler's differentiation between communities of feeling and fellow-feeling. In communities of feeling, for example, grief is shared for the loss of a jointly beloved person: 'fellow-feeling would be when I feel sorrow about your grief although I do not share your object of grief … your grief is what grieves me; your grief is the object of my grief' (Ahmed, 2010, p. 57).

The sharing of grief might also help us avoid the solipsistic reduction and sanitising of queer grief into and as an *object* – a material thing to be passed hand to hand that absolves majoritarian society of the need to grieve queer lives as a community – for example in the form of the NAMES project AIDS Memorial Quilt (Crimp, 2002). Rather, objects made by and out of grief clear a space for sharing and grieving those we have lost without making memories and feelings of grief into objects that can be taken up or appropriated or put on display as just another memorial produced by the nation state (Plummer, 1995; Ahmed, 2014). Ahmed reminds us, 'Queer activism has consequently been bound up with the politics of grief, with the question of what losses are counted as grievable' (2013, p. 164). She draws on their theorisation of how AIDS transformed the notion of public grieving, and collective loss. But, importantly, she asks the question: 'what are the political effects of contesting the failure to recognise queer loss by displaying that loss?' (2014, p. 188).

In this way, *things* such as the AIDS Memorial Quilt are not so much objects *as* memorial as they are objects *becoming* memorial as they stitch together new ways of relating. As an example, Ahmed writes of George Eliot's *Silas Marner* and the loss of an earthenware pot – a loved object he used to fetch water for years. When the pot breaks, its use and Silas's way of relating with it shift;

though what remains in the shattered pieces is the affection of the relation. With the breaking – and the loss – a new way of relating is created – the pot is no longer an object but a process of 'becoming memorial, a holder of memories' (Ahmed, 2014, p. 45).

Of the shift from the object to the relation of queer grief Ahmed writes:

> a queer politics of grief needs to allow others, those whose losses are not recognised by the nation, to have the space and time to grieve, rather than grieving for those others, or even asking 'the nation' to grieve for them. In such a politics, recognition does still matter, not of the other's grief, but of the other as a griever, as the subject rather than the object of grief, a subject that is not alone in its grief, since grief is both about and directed to others.
>
> *(2013, p. 44)*

To be not alone in our grief. Something seemingly so fundamental and simple, so often brushed away. Though if grief is a language – and a body – doesn't story, doesn't song, open up space and a time to grieve?

I sat by her bed and read to her. My mother was a deep lover of words. I'd just started my PhD and the only book I had with me was *Pedagogy of the Oppressed*, by Paulo Freire. I tried to read it to her, but she brushed it away. It was confusing her, as her brain function lessened. I closed the book and told her stories instead. Stories from my childhood, things we had done as a family. Anything I could think of. I made up stories too: about a little girl who goes for a walk in the forest, finds magical animals and purple trees, a bright red carousel. My mother loved it. She would squeeze my hand when I made up a detail that she particularly liked.

Eventually, though, she stopped responding and slipped further into unconsciousness. So I sang to her instead, all the old songs she loved: spirituals, Neil Diamond songs, children's rhymes, music as a holder of memories. I stroked her arms and hands and face, her feet and hair, and I sang.

A Liveable Life, a Grievable Death

So many of us are 'at sea' in the presence of loss in the absence of social supports and conventions that make grief recognisable and thus inexpressible in majoritarian society. This is one of the reasons why 'same-sex marriage' is so important not only legally but also symbolically and relationally. It is also one of the sites of evidence that changing laws is not the same as changing *custom* or making possible the complex sharing of emotion around the experience of pledging our commitment to others or mourning their passing. Not everyone recognises gay marriage; not everyone respects it. It remains unintelligible in some places, illegal in others. It also serves the dual dangerous function of allowing majoritarian culture to think that everything is 'equal' or resolved now, in terms of queer grief and queer love, while giving queer persons a false sense of security that we are somehow 'equal'. That is, even now that 'marriage equality' has passed in so many western nations, it is still a precarious and sometimes unintelligible kinship. And of course, legal and legislative progress is not a mono-directional project and is always contingent upon the *mores* of each time.

The absence of convention leaves most people anxious about how to address grief, so most often they/we don't (Jurecic, 2015). But there's more. As a socially inscribed, culturally constructed entanglement in which the grieving emerges from within the event, all mourning is not performed equally. Grief, as Ahmed points out, has a sociality that is heteronormative. When we become entangled in queer relationships, we leave behind accepted social paths and support systems of protecting and nurturing collectives and communities (Ahmed, 2016, par. 39). Compulsory heterosexuality is an 'elaborate support system – the path is kept clear to ease a progress, loves cherished, losses mourned' (Ahmed, 2016, par. 39). This lack of elaborate support system is one aspect of the unintelligibility of queer grief. Loss is heteronormatively inscribed, and so queer loss is often the place of no rules, of the unknown, of heteronormative kin (and strangers) not knowing how to respond, even if they see or are made to see, the kinship. Heterosexual kinship networks and heterosexuality itself become

> a form of having…. The history of heterosexuality, we might even say, is the history of broken hearts, or even just the history of hearts…. With such recognition, comes care, comfort, support. Without recognition, even one's grief cannot be supported or held by the kindness of another.
>
> *(Ahmed, 2010, pp. 108–109)*

Recognition, then, is the linchpin connecting not only love and grief, but also the politics of being seen as grieving, as having a right to grieve, and as making possible the support and kindness of a community afforded to those who are grieving. Judith Butler's work (2005, 2006, 2009) on what makes a life grievable (and thus what makes life itself possible), teaches us about the politics of recognition, which work within 'normative schemes of intelligibility [that] establish what will and will not be human, what will be a livable life, what will be a grievable death' (p. 146). These normative schemes grow up out of the social relationships which we understand as places of nurturing, care and community, such as the family and school (Harris, Holman Jones, Faulkner & Brook, 2017). These are also the very places and relationships where we learn and unlearn (through naturalisation) notions of exceptionalism, individualism and hierarchy and the practices of exclusion that support them. As Butler (2006) puts it, 'if a life is not grievable, it is not quite a life; it does not qualify as a life and is not worth a note. It is already the unburied, if not the unburiable' (p. 34).

The politics of intelligibility and recognition depend on normative practices; yet these same practices are also upended by stopping long enough to recognise how any notion of 'us' or 'we' depends on the recognition of the other. Thus, recognition is also how we 'confound identity' and see each other other-wise. Butler writes:

> For if I am confounded by you, then you are already of me, and I am nowhere without you. I cannot muster the 'we' except by finding the way in which I am tied to 'you,' by trying to translate but finding that my own language must break up and yield if I am to know you. You are what I gain through this disorientation and loss. This is how the human comes into being, again and again, as that which we have yet to know.
>
> *(2006, p. 49)*

I am nowhere without you. I am tied to you. And if I am to know you, my own language must break up. Yield to disorientation and loss. Recognising not only our dependence on one another but also our need to yield to loss makes possible a way for language – however broken, however disoriented – to answer Ahmed's question: what would it mean for the 'ungrieved to be grieved' (Ahmed, 2013, p. 176)?

Reading Grief, Writing Melancholia as an Archive of Survival

The doctor told us she would be gone in less than 72 hours, but she lasted 11 days.

They moved her upstairs to a room, and my brother and I stayed with her, sleeping fitfully in the room next to her bed.

Bright and early the next morning, a young physical therapist chirped into the room.

'Good morning Mrs Harris!'

Mark and I jumped.

The therapist shouted into my mother's still-blinking face that she was going to get her up and moving, in no time.

We were confused. I could only imagine what my mother was thinking.

The therapist put an ice chip into my mother's mouth. That's what they do for stroke patients when they lose control of the muscles in their mouth, their throat, and can no longer swallow. The ice cube melts on her tongue and keeps her lips and mouth lubricated without sips of water, which only gurgled down her chest.

My brother tried to whisper when he said, 'she is not going to get up and moving, ever.'

The physical therapist was committed to her job. 'Oh, we can get miracles of this little lady!' she said.

Mark and I took her out into the hallway and after trying momentarily to be polite, he shouted, "Didn't you even read her goddamned chart? She's going to die!"

But the emergency room chart had not made it upstairs with her to the room.

Trying to pass the long anxious grief-filled nights, I'd sit in the chair beside her bed, and try to finish *Pedagogy of the Oppressed*. I loved it in principle.

So many of my friends and heroes claimed that it completely transformed their lives.

But the words were like straw in my mouth.

Even after she died, and I worked on that thesis for two more years, I never could go back to that book.

And I find it hard, even now.

Sally Munt argues that

> the anger produced by grief enables the subject to force a separation which ultimately, when it has played itself out, enables the subject to recover a coherent sense of selfhood. If mourning is to be successful, the anger will be replaced by a true acknowledgment of loss including a wish for repara-tion…. Narratives become the performative force of grief, they represent the moment of death and they revise it, replay it, until the memory becomes de-cathected. They bring the two temporalities together – the original event, and our grieving present, and they can *move* us to restitu-tion, through the passages of time that reading, and grief, requires.
>
> *(2007, p. 135)*

Sigmund Freud said that melancholy is the twin of mourning, except for its pathological persistence and self-disregard. Though if 'I' am nothing without 'you,' then my self – my coming into being, even after you're gone – isn't coherent, or de-cathected. The loss of you isn't repaired or recompensed, no. And time and reading don't make for transformation or resolution. What might we make of narratives that perform the 'performative force of grief', then? Perhaps we might, instead, use writing about loss as a reminder that the promise of love always contains, ultimately, the promise of grief. That to avoid grief is to avoid love altogether:

> Sadness the blue like dusk, the reminder that all things are ephemeral, and that because there is time there is change and that another name for change, if you look back toward what is vanishing in the distance, is loss. But sadness is also beautiful, maybe because it rings so true and goes so deep, because it is about distances in our lives, the things we lose, the

abyss between what the lover and the beloved want and imagine and understand that may widen to become unbridgeable at any moment, the distance between the hope at the outset and the eventual outcome, the journeys we have to travel, including the last one out of being and on past becoming into the unimaginable.

(Solnit, 2013, p. 261)

When loss comes, mourning and grief become an 'expression of love ... love has an intimate relation to grief not only through how the subject responds to the lost object, but also by what losses get admitted as losses in the first place' (Ahmed, 2010, p. 139). Queering loss as an expression of love between language and bodies and books: allowing the ungrieved to be grieved.

My father had a second heart attack when I was doing my PhD at the University of Texas; he was treated with a stent in the outpatient facility, so my mother called me after he'd already been sent home. But when he had his third, and again needed open heart surgery, my mother called and again I went. I left my desk piled high with materials I was assembling for my tenure dossier and boarded the plane.

This time my waiting room book is DJ Waldie's *Holy Land*, a memoir I'd read and loved in graduate school and had now assigned to my own students. The book is in part about Waldie's struggle to balance his father's ideas about the world and what makes a livable – and thus grievable – life with his own. It calmed the panic that I felt waiting for my father to come out of surgery; the sadness I felt when I spoke to my father about my own coming out. It put word and force around the grief we must all feel when our relationships with our parents change – when we change and we can no longer persist in seeing each other for how and what we were, as we were before (Butler, 2005). When the only real forces at work between you as you come to terms with grief are circumstance and grace.

I sit next to my father's bed in the recovery room. I ask him if he'd like to hear what I'm reading. He whispers a hoarse 'yes.' I read: 'He could not choose to deny his father, even less his father's beliefs. These have become as material to him as the stucco-over-chicken-wire from which these houses are made' (Waldie, 1997). My father nods and closes his eyes. He sleeps and I stand, book in hand, and move into the hallway to stretch my legs. When I return to my father's room, the nurse is at the door.

'Can I sit with him?'

'In a moment. I'm afraid, though, that there's been a change in your father's condition.'

'Changed? I was just in with him.'

'He's had a stroke. I'm not sure how significant. They're going to take him downstairs for some tests.'

I ask to see my father and the nurse says she'll ask. While I wait, I read from *Holy Land* aloud, to no one in particular: 'At some point in your story grief presents itself' (Waldie, 1997).

Michele Pearson Clarke (2017) writes about the more-than-sadness of her grief in losing her mother, and its aftermath. While her mother had come to terms with her cancer and imminent death, Clarke could not. It opened up a place of generative but devastating 'before-and-after', what she writes as the failure of her queerness to shield her from other life blows. Clarke reminds us through her work that grief, like oppression, is intersectional. She reminds us that death and mourning, like queerness and gender nonconformity, remain taboo. Myles Mason (2017, n.p.) writes about 'queering the affective potential of grief', drawing on Raymond William's structures of feeling. And Catriona Mortimer-Sandilands (2010, p. 31) understands grief as a political condition, writing:

> Melancholia suggests a non-normalizing relationship to the past and the world, in which the *recognition* of the identificatory persistence of loss in the present – loss *as* self, the fact that we are constituted by prohibition, power, and violence – is central to our ethical and political relationships with others. Or, as Butler writes, grief "furnishes a sense of political community … by bringing to the fore the relational ties that have implications for theorizing fundamental dependency and ethical responsibility.

So many have written about the social, corporeal, and political conditions in which these ethical relationships have been revealed with particular force and clarity to the queer communities and politics, particularly as they were ignited and galvanised by AIDS and the defiant, rage-filled grief (see, e.g., Butler, 1990; Phelan, 1997; Crimp, 2002; Cvetkovich, 2003; Schulman, 2012). For Cvetkovich, the melancholic insistence on archiving the trauma of the past is an embodied and 'de-pathologizing' process for understanding how grief, rage and loss are 'felt experiences that can be mobilized in a range of directions, including the construction of cultures and publics' (2003, p. 47). The 'public melancholy' of loss, for queer people, is a 'form of survival' (p. 342). Could it also become a form of survival for not only human-kind but also other animals, plants and perhaps our dying planet?

Grief and the Anthropocene

The writing retreat, and with it the silver-green-grey leaves, the misty mornings and the sun-dappled afternoons, and the sulphur-crested cockatoos, are gone. Not gone-gone, but passed into memory as a week and world away. Returning to our writing on bodies and languages of grief, I am taken back to a walk on our last morning in the country, along a walking track that wasn't. It was, instead, a track for accessing the electrical poles rising up out of the landscape, their coarse straightness out of place in the swaying forest of mountain ash. I think back to that walk, to the way my attention was drawn by and into the soft moss floor of the forest, looking out for branches and brambles strewn across my path. A path made by trespass in the name of human need and entitlement. A path made by or because we are now living in the Anthropocene – the geological epoch named to mark human-kind's lasting and devastating impact on the planet.

Back at home (or is it?), I look at photographs of the Toolangi State Forest, home to giant eucalyptus trees, habitat to the threatened Leadbeater's opossum and 19 other rare and threatened plant and animal species, diminished and driven to the point of extinction by decades of state-sanctioned logging and devastating bushfires. I think about how we retreated to the forest with human companions to make space outside the world of neoliberal demands while forgetting, perhaps, how the forest itself breathes and grows inside the world of neoliberal demands. It is both a church in nature and a place in need of sanctuary. I think of how this essay on grief might also write of nature and all the species in it – human bodies, animal bodies, plant bodies, thing bodies – as bodies in relation and how then we might come closer to understanding grief and melancholia as a kind of recognition. A way of a allowing the ungrieved to be grieved and as a form of and perhaps for survival.

Richard Anderson (2001) says that grief is the heart of the modern age and human ravaging of the planet. He writes:

> At some level, we're aware that something terrible is happening, that we humans are laying waste to our natural inheritance. A great sorrow arises as we witness the changes in the atmosphere, the waste of resources and the consequent pollution, the ongoing deforestation and destruction of fisheries, the rapidly spreading deserts, and the mass extinction of species.... It's necessary to face our fear and our pain, and to go through the process of grieving, because the alternative is a sorrow deeper still: the loss of meaning. To live authentically in this time, we must allow ourselves to feel the magnitude of our human predicament.

Feeling the magnitude of our predicament – human and more-than-human – is what Donna Haraway has been asking us to do for a long, long time now. Haraway's (2008) notion of companion species is a way of understanding not only how we are all beings formed in the co-shaping entanglement with others but also how we must come to terms with such interspecies relationality in order to survive. Predicated on the very notion of 'I' am nowhere without 'you' – human, plant, animal and other – 'species interdependence is the name of the worlding game on earth and that game must be one of response and respect. That is the play of companion species learning to pay attention' (p. 19).

Paying attention demands that we recognise that we do not, actually, make something out of nothing; rather, we are who we 'become with companion species, who and which make a mess out of categories in the making of kin and kind' (Haraway, 2008, p. 19). Disorientation. Yielding to loss. Loving and grieving in an ongoing, melancholic (and de-pathologising and perhaps decolonising) bid for survival. Breaking open language and recognising not only our dependence on one another but also how 'grief reworks truth to tell another truth' (Haraway 2008, p. 178).

References

Ahmed, S. (2016). Wilful stones. *Feministkilljoys*, 29 January. Online: https://feminist killjoys.com/2016/01/29/willful-stones/.
Ahmed, S. (2014). *Willful subjects*. Durham, NC: Duke University Press.
Ahmed, S. (2013). *Cultural politics of emotion*. London: Routledge.
Ahmed, S. (2010). *The promise of happiness*. Durham, NC: Duke University Press.
Anderson, R. (2001). The world is dying – and so are you. Commentary. *Los Angeles Times* 7 January. http://articles.latimes.com/2001/jan/07/opinion/op-9312.
Butler, J. (2009). *Frames of war: When is life grievable?* London/NY: Verso.
Butler, J. (2006). *Precarious life: The powers of mourning and violence*. London/NY: Verso.
Butler, J. (2005). *Giving an account of oneself*. New York: Fordham University Press.
Butler, J. (1990). *Gender trouble: Feminism and the subversion of identity*. New York/ London: Routledge.
Clarke, M. P. (2017). Parade of champions: The failure of black queer grief. *Transition*, *124*(1), 91–98.
Crimp, D. (2002). *Melancholia and moralism: Essays on AIDS and queer politics*. Cambridge, MA: MIT Press.
Cvetkovich, A. (2003). *An archive of feelings: Trauma, sexuality and lesbian public cultures*. Durham, NC: Duke University Press.
Haraway, D. (2008). *When species meet*. Minneapolis: University of Minnesota Press.
Harris, A. M. & Holman Jones, S. (2018). *Queering autoethnography*. NY/London: Routledge.
Harris, A. M., Holman Jones, S. Faulkner, S. & Brook, E. (2017). *Queering families, schooling publics: Keywords*. New York: Routledge.
Jurecic, Ann. (2015). The art of medicine: No protocol for grief. *The Lancet, 386*, 848–849.
Macdonald, Helen. (2014). *H is for hawk*. London: Vintage Books.
Mason, M. (2017). Good mourning: Structured feelings and queering the affective potential of grief. Unpublished doctoral thesis.
Mortimer-Sandilands, C. (2010). Melancholy natures, queer ecologies. In C. Mortimer-Sandilands and B. Erickson (Eds.), *Queer ecologies: Sex, nature, politics, desire* (pp. 331–358). Bloomington, IN: University of Indiana Press.
Munt, S. R. (2007). *Queer attachments: The cultural politics of shame*. Abingdon/NY: Routledge.
Phelan, P. (1997). *Mourning sex: Performing public memories*. New York/London: Routledge.
Plummer, K. (1995). *Telling sexual stories*. London: Routledge
Schulman, S. (2012). *The gentrification of the mind: Witness to a lost imagination*. Berkeley/ Los Angeles: University of California Press.
Solnit R. (2013). *The faraway nearby*. New York: Viking Penguin.
Waldie, D. J. (1997) *Holy land*. New York: St. Martin's Press.

2

"I WANTED THE WORLD TO SEE"

Black Feminist Performance Auto/Ethnography

Wilson Okello

July 13, 2013: I recall the scent and sensibility of linen incense burning late into the evening, its aromatic opulence lingering as if it too was bracing itself for the news to come. Smoke raised itself from the controlled flame like ancestors coming to bear witness, to signal devastation or cast warning in the belief of rainbows, either way there was a message on the way and I/we should start preparing ourselves for it. I pressed my eyes on CNN, with distrust and every bit of hope I could dig up.

The day itself seemed to be dressed in fatigue, bearing a weight just beyond its ability to carry. We have seen this song and dance before. It is the type of exercise that made us strong; the type of longsuffering that is borne in resilient imagination. Justice is judged differently in my world. I learned that I was scary, that my body was threatening, before I could articulate love. When you pose a danger to someone in my world, there is the chance that you could be punished for it, consequences of which may include your life. Protect your body at all costs. Protect the lives of your brother and sister at all cost. These maxims leafed with others, constituted the talk, a survival manual issued by my parents and enforced by streetlight regulations are still vivid memories, shifting, and repositioning themselves in my consciousness over time, always present.

This account is a reaching back; I am seeking to recover the memory of a boy ... a boy who could have been me. At a different moment in time, he was me. The death of Trayvon Martin stirred something up in my innermost parts. Fear, most notably, a type of haunting that has echoed into sleepless nights and desperate attempts to make right the world's collaps-ed/ing virtue. Too, it has meant purpose toward a vocation and a decision to never, again, *walk in fear*. I have spent the years following this catastrophic and crucible moment demonstrating through performance poetry that the Black body is *capable, and worthy, of testimony every time it shows up in a public space.*

Black Bodies in the Wake[1]

Where there is breath, and breathing, there is the opportunity for life and resistance. Living, as Bailly and Anderson (2010) claim, "is porous, intimating its
capacity to be an antidote against enclosure, seizure, capture, incarceration,
enslavement" (pp. 4–5). White supremacy, a panoptic controlling instrument
toward and against Black bodies in and across the diaspora, maintains its rational,
seemingly objective authority, by schooling the Black body. Schooling, as Hill
(personal communication, February 1, 2018) notes, are the curricular and con-
ditioning strategies that function to reproduce the status quo. Whiteness, as a
result, becomes the fundamental organizing principle that structures Western
society, and imperialism globally, grounded in the need to explain and define
the boundaries of existence.

Baldwin (1985) supports this notion as it pertains to Black expression, writing
"the black man has functioned in the white man's world as a fixed star, as an
immovable pillar: and as he moves out of his place, heaven and earth are shaken
to their foundations" (p. 336). The expectation communicated here, is that
Black bodies make peace with the limits of the white imagination. For the
Black body, this has meant *stay in your place*, or be *put in your place*. Baldwin's
metaphor of a star—an extraordinary incantation of great power, pull, and
mystery—is a useful illustration here, because Black bodies have always been as
fascinating as they are overwhelming in the white imagination, and thus, a thing
that can and should be controlled (hooks, 1990). In instances where the Black
body misbehaves, breaks with the aforementioned project of schooling,
dominant forces spare no effort to reassert their dominance, by stilling, that is,
the act of making something motionless, static, or inanimate—the Black body.
History owns a long record of terrors against the Black body in an effort to
remind Black life-worlds of their fixed positions—Black people receiving less
medical treatment, because they are thought to feel less pain; enforced steriliza-
tions of Black women; mass incarceration—a concept that Sharpe (2016) calls
the wake.

In articulating the wake, Sharpe (2016) describes it as "the history and
present terror, from slavery to the present, as the grounds of everyday Black
existence; living the historically and geographically dis/continuous but always
present and endlessly reinvigorated brutality in, and on, our bodies" (p. 15). In
what follows, I ponder what it might mean to reclaim what Spillers (1987)
describes as "the flesh and blood that was essentially ejected from the ... body
in western culture" (p. 67), toward an embodiment, or the merging of subject-
object positions. Making a distinction between the body and the flesh, Spillers
(1987) suggests that "before the body there is the flesh" (p. 67); a difference
between captive and liberated subject positions, such that the body before it was
inscribed and imposed upon by the discourses of rationality, is flesh—a creative
entity and grounds by which life occurs. To clarify this point, I turn to the

genius of Black mothers, who have routinely engaged in public, self-defining praxis (Okello, 2018), as a way to make meaning of their lives in the wake (Sharpe, 2016).

Limbos in the Wake: A Black Feminist Performance Auto/ Ethnographic Approach[2]

Sybrina Fulton, Geneva Reed-Veal. These mothers, and the many others not named, were charged with the pressing question: what do I do with the body of my child? Here, I consider the body as many things, including history, memory, and legacy. Invoking the wake, as described above by Sharpe (2016), what viewing, opportunities for mourning or righteous indignation, should these mothers have turned to alongside the wake of history that befell their children? With limited choices available to them, they all appealed to the spirit of Mamie Till-Mobley, a mother like them, whose flesh gave breath to a lifeless body. Till-Mobley taught us about the educative/ing body in the wake (Sharpe, 2016), facilitated by the stillness of her son. The insistence on fixedness or what I term *stillness*, can result in a socially sanctioned array of counter moves against the perceived source of the discomfort, including: penalization and retaliation, or what DiAngelo (2011) calls fragility. More aptly, the Black scene in the white imagination is understood as a phobogenic object (Gordon, 1997) occasioning anxiety and stress, which warrants the repression of the distressing object under the terms of rational policies and practices. Emmett Till, allegedly, placed his tongue on the roof of his mouth, filled his lungs with air and blew, engaging in the double gesture of inhalation and exaltation. Emmett's crime was one of Black breath, an offense intimated against whiteness—the threat of desire outside of the apportioned lot of Black ontology (Kendi, 2017)—that would ultimately still his body.

With radical clarity, at the funeral of her son, Mamie Till-Mobley proclaimed, "I wanted the world to see what they did to my baby" (Robson, 2010, p. 26). Till-Mobley insisted on providing the undoubtedly painful and repeatedly traumatizing visual of her broken son as the ongoing manifestation of historical terror. She does this while also providing images of a smiling, whole Emmett on and around his casket. This careful and difficult decision assists me in remembering Emmett, and the many other slain and broken bodies (spirits, souls, and voices) that have been swept up in the wake (Sharpe, 2016).

Callier (2018) calls upon M. Jacqui Alexander (2005) and Alexis Pauline Gumbs (2014) to denote palimpsest as "time and history collapsed, rewritten upon itself, effaced, and revealing in the present time glimpses of a still/here, present past" (p. 17). Black bodies in motion, stillness, and the conscious and unconscious performative acts that seek to confer life and legibility onto minoritized bodies inspire this methodological approach. The body is more than just the here and now, but is an archive of memories, histories and dreams, as well as the probable violence, epistemic and material, that is waged against them as a consequence of living from day to day. Invoking Mamie Till-Mobley's lamentation of the wake (Sharpe, 2016) in "wanting the world to see," I take up the full weight of the historical and biographical body and weave it together with Joy James' (1999) notion of *limbos*, to respond to the question: *how might centering the body clarify possibilities of liminality in the current sociopolitical order?*

Choreographing Limbos

Limbos are known to entail "vulnerable backbreaking postures as well as isolated states" (James, 1999, p. 41). In these limbos, Black feminisms cultivated models and strategies for resistance that rejected injunctions of the status quo. According to James (1999), limbo, in its primary usage, references liminal spaces, "oblivion and neglect, or suspension between states" (p. 42). Its secondary, and more playful usage, is characterized by "play, struggle and pleasure— the black/Caribbean performance where dancers lean backward, with knees bent to pass below a bar that blocks their path. It represents determined progress despite the vulnerability of the position" (p. 42). As choreographers of political agency, Black feminisms' conservative or liberal ideologies conduct varying limbos as they negotiate the plurality of their meanings. They advance emancipation projects while simultaneously being distanced from the center of issues of oppression, and yet, continuously extract agency from these margins. This fluidity allows them to interpret, dissect, improvise, describe and agitate, displaying an agility and imaginative power to deconstruct and reconstruct. James (1999) conveys the work of such limbos as that which synthesizes emancipation theories. They assist in illustrating and analyzing the intersections/multidimensionality of oppression and freedom. Moreover, limbos are ideal spaces for *witnessing*, giving language to experiences, which might otherwise remain muted. Finally, in their progressive movement forward, limbos "often bend backward toward historical exemplars to retrieve from the sidelines of conventional memory important ancestral leaders for current considerations and political struggles" (James, 1999, p. 43). In this way, limbos are performative enactments. The following section details this connection.

Limbos as Performance

Performance, according to Bell (2008) "is a way of knowing ... a way of staking claims about the creation of knowledge" (p. 18). Given the improvisational, ephemeral, emotional, and physical nature of limbos (James, 1999), one should constitute their deployment as performance—both process and product—or that which demonstrates the connections, extant or otherwise, between self, other, and culture. Limbos as performances engender a richer understanding of who one is in relation to others in culture (Spry, 2011). Concerning process, as an analytic method, performance assists in materializing fissures in the status quo, giving a face and body to abstract conceptualizations of violence. Additionally, it is an intergenerational stained grounds, evidencing the perpetual vulnerability to premature death enveloped in *stamped* (Kendi, 2017) bodies. As a product, which constitutes here the act of creating, limbos signal to readers/viewers a breathing portrait of possibility, predicated upon the necessity of ongoing self-defining praxis (Okello, 2018) in the face of schooling procedures that function to still the body. Refusing to allow stillness to have the final word, Till-Mobley did not seek out rational methods for translating her experience; rather, she sought marronage—a freedom from the closure of stillness; a choice to speak of life and love when death sought to close her, and Emmett, in.

If schooling (Hill, 2018, personal communication) marks the interdiction toward a civic normalcy that depends on the explicit and unquestioned unperceivability of Black bodies, performance is rupture. Excess motion, breath, speech, writings, and other signs of life are and have historically been the makings of offensive behaviors due to their intuitions of efficacy. As a retort to schooling and stillness, history cites exemplars who engaged in limbo (various cultural performances). In the post bellum era, for example, Ida B. Wells-Barnett published the anti-lynching pamphlets, *Southern Horrors: Lynch Law in All Its Phases* and *A Red Record*. These performances of excess—of life against stillness—proclaimed to "give the world a true, unvarnished account of the causes of lynch law in the South" (Gabbidon, Greene, & Young, 2002, p. 25). Her campaign, like that of the Black mothers mentioned above, enlarges our understandings of performance as they keyed into the how—how do lynchings occur?—and the why—why do lynchings occur?—of the very phenomena under investigation. Performing limbos, Wells-Barnett stated:

> The purpose of the pages which follow shall be to give the record which has been made, not by colored men, but that which is the result of compilations made by white men, of reports sent over the civilized world by white men in the South. Out of their own mouths shall the murderers be condemned.
>
> *(in Gabbidon, Greene, & Young, 2002, p. 64)*

With intentioned pen, Wells-Barnett detailed the realities of lynching throughout *A Red Record*.

In *Southern Horrors* she illustrated the *why* as tied to economics and political motivations. According to Muhammad (2010), "professional and entrepreneurial blacks were frequent targets of mob violence in the South, especially when their commercial activities weakened the grip of white business owners who systematically exploited blacks" (p. 59). Crawley (2013), linking breath to the assumed, undiluted, purity of White inhalation and exhalation, states: "Breathing, to be blunt, shares a relation with any political economy. Lynching, as a peculiar performance of moral uplift for whiteness and moral degradation against blackness, was grounded in concerns over economy" (p. 58). Having borne witness to the rationalized justifications for lynching (as the most effective way to manage Black criminality) by white journalists, Wells-Barnett wielded these performances as a way to speak truth to power. More than a solo cause, these limbos were animated by a desire for others to experience life anew in and against threats of stillness.

Performance auto/ethnography works with(in) a sociopolitical commons, that is, a specific cultural context to explore bodily knowing, and thus, makes use of and complicates what is most familiar in an effort to extrapolate the plentitude of meanings situated on its grounds (in the case of the performance auto/ethnography, the body). Relinquishing the need to arrive at one, common, singular experience to reflect the identity of stamped (Kendi, 2017) bodies in the wake (Sharpe, 2016), a performance approach can open constructs of race, gender, class, sexuality, and ability to greater vitality, variety, ambiguity, and possibility (Jones, 2002). As such, the following question emerges for this methodological approach: what determines the capacity of any one body to *do* limbo?

McKittrick (2006) considers the Black body as both a *site* of pain and a *sight* of possibility. In conversation with Diawara's (2011) descriptions of the diaspora as "the passage from unity to multiplicity," performance is a reminder that "the multiple is a fact of the unit, that within any object are multitudes" (Crawley, 2013, p. 50). Variations are integral to the reconstitution of that which is at once denigrated and, too, a source of critical distancing from penalizing scripts. Employing these concepts, Leigh Raiford (2006) states the following about lynching photography:

> [W]e can understand these images to be sites of struggle over the meaning and possession of the black body between white and black Americans, about the ability to make and unmake racial identity. In the hands of whites, photographs of lynchings, circulated as postcards in this period, served to extend and redefine the boundaries of white community beyond localities in which lynchings occurred to a larger "imagined community." In the hands of blacks during the same time period, these photographs were recast as a call to arms against a seeming never ending tide of violent coercion, and transformed into tools for the making of a new African American national identity.
>
> *(p. 22)*

Recasting the same material in the circulation of photography is to aestheticize it, to generate new meaning.

The intellectual-made-corporeal performance of limbos, and dispersal of affect through the aestheticizing comport of Black limbs (James, 1999), serves as a critique of schooling and anti-blackness, and an enunciation of desire to openly participate in more expansive qualities of life, even as one is situated within scenes of subjection (Hartman, 1997; Scott, 1990). I return to the body in and through performance to make recorded histories and lived realities in the wake (Sharpe, 2016) of our experience (Jones, 1997). I return to generate a narrative about "the intentioned forms black life-worlds took, and take, to contend with and against a history of violent eclipses of breath—drownings overboard, whippings, lynchings, incarceration—and how those forms of life are the instantiation" (Crawley, 2013, p. 57), and incarnation of life and existence in the wake (Sharpe, 2016) of gratuitous violence and violation; and as Jones (1997) instructs, I return to and examine the body to answer the embedded question of every performance: what are the theoretical implications of doing limbo in the wake (Sharpe, 2016)? This question is anchored in the belief that every performance is distinct, constituting a new arrival rather than a repetition of sounds and movement.

Limbos as performance have epistemic and ontological relevance. To get at these implications, limbos are choreographed in step with what Aisha Durham (2014) organizes as recall, (re)membering, and representation. In what follows, I offer my poem "Buried" and explain how Durham's procedures buttress this methodological approach.

Buried

I hear it/it's faint
ear to your chest, drifting
but I hear it
this world doesn't expect you to make it;
warnings littered in the church hallways and sunday school lessons/
sanctuary pieces cover our peace while we are still in it/
buried/in rubble/in Birmingham's, Audubon's, balcony's,
and pavements as headstones
oceans roar with our echo
these
cemeteries
are tight
too small
too soon
I'd jump too
Lay me to down to sleep
So I bury it and stand watch
before we ever step on stage/stand up in rage
bury it
before it's taken/sentenced/stolen/persuaded away,
I bury it
deeper/beneath dirt plot/before wakes and eulogies/
kill the thought
so I won't believe it—or dig it up like should I be here

Like/I know things/
Like, I should be here and know things
Don't nobody want you—I know it
Don't you think I know it?
I don't blame you/how you crawl into yourself
Close and lock yourself in,
It's dangerous
Beyond these walls/outside of the jaws
where you have learned that your words are laid to rest before they are born
where every letter you put together is more than anyone thought you could utter
sit down/stay put/lie still/that mic ain't magic—something don't come out of
nothing
but
I hear it/it's faint
ear to chest, drifting

I see it/rise and fall compressions
There are hidden things in me too
Hushed and mute and tired/fit to be tied
I know you are tired/of explaining/of forgiving/
/of needing and nobody askin
heavy hearts and hollow points
victim blaming and niggas just doing their thing,
I get if you are tired of bein powerful
I know—this life can be its own coffin
Embalmed but withering until nothing remains
But where there are bones
and skeletons
there is more to life than just dying daily
And I've salvaged the bones
where there was voice and body
ashes and dust
I believe in resurrections
Dead things.
Buried and bound come back to life.
But this stone
This stone
Is
Heavy
And I can't do it on my own
Lift it on my own
Get up/stand in/or up
Return to me

Give it back
Return us to ourselves
We are stolen
Give us back
[The house of bones]
bring me home.
To myself. Bring us out
And home
I'm tired
Hold her up/she's weary
My arms are tired
Stand in
On the dome of my mouth
They are weeping
And my arms are weary
Get up/like you said
Get up like you said you would
You can do it
Muscle and bone and prove the doubters wrong
Don't touch her
Don't touch him
Don't—don't nobody touch them—they are dreaming
I knew them
Rocked and named them
Get up/
Sleeping giant/shotgun/
Brewing tornado
I don't care
I don't care that nobody wants you
You are mine
Return to me
Rose and beanstack—[wish to live]
I want to live and can't do it without you
There is fire
And—And rain
Tsunami and flowing volcano
eden and eruptions—
You harness both

tell me what you need
and I'll get it/I'll do it
but if this it, let me pay my respects/
chisel on your tomb/buried here are miracles/black ones/

surviving lifetimes/warrior who would not die when buried alive
here rests a psalm—raw/and rage she is rhythm and sorrow in hallelujah chants
an ode to complexity in all its beauty
worthy of time to her self
whatever you choose/let it be/for you

Recalling Archives

Before a theory through the body can be offered, one must engage in *recall* to show that the auto/ethnographic *I* comes from somewhere. Spry (2011) denotes the performative *I* as a critically reflective disposition that accounts for one's multiple and ever-changing identities. Further, she contends that "it is a negotiation of representation with others in always emergent, contingent, and power-laden contexts" (p. 54); recall insists that I weave an awareness of my positions into and onto the body of the performance. Pulling from their autobiographical basins makes the researcher acutely aware of how they witness their reality and the ways in which it has been pieced together (Durham, 2014; Hill, 2014; Spry, 2001). As it relates to the theorizing coming to bear in this performance, recall provides context, by showing *what I was made to be.* Operating from this standpoint pushes constructivist subjectivity (Baxter Magolda, 2009), or the notion that all people are equipped with the sociopolitical power to define themselves, to a place of criticality as it forces one to contend with history and historicity, stories and assumptions, the ephemerality of Black bodies always in motion across time.

This poetic representation drew from the well of memories I stored regarding Trayvon Martin's death in 2012. Particularly, I invoked the ethnographic *I*, that is, the researcher who is able to see their body as more than just a template to be written on by the world but as "instrumental to sensing the gaps, holes, fissures, and fixtures of culture and identity" (Durham, 2014, p. 22). Memories excavated for analysis are only as useful as the interpretive criteria employed to make sense of the archival materials (Hill, personal communication, November 1, 2017). As I brought myself into the archive, lifted up the corners, shook and dusted off the artifacts, I endeavored to make meaning of what was uncovered by asking several interpretive questions. This archival work made use of memories by responding to the following: *What is the work of memory?* This question asked if the memories being invoked brought to life, through the body, the larger phenomena being examined. In other words, for what purposes am I engaging with the memory of Martin's murder and what do I want to accomplish? Above, I note this rationale as, "purpose toward a vocation and a decision to never, again, *walk in fear.*"

Second, I asked, *What is the feel of memory?* This question held me accountable to the body as a way of knowing through description and contextualized materials. Pointedly, what did I feel as I engaged the memory? Where did I feel it? In the poem, this sentiment comes through when I write *"My arms are tired/ Stand in/On the dome of my mouth/They are weeping/And my arms are weary."* Finally, I must ask, *What is the meaning made of (re)hashing memories?* Ascertaining meaning must take into account prior understandings, encompass all that is relevant about the phenomenon, and ultimately leave room for further analysis, as all interpretations are unfinished (Denzin, 1994; Hill, personal communication, November 1, 2017). In this sense, I maintain that Trayvon Martin's premature death has something to teach the public about living in the wake (Sharpe, 2016). Despite the desire for whiteness to erode Black life, it persists. Black bodies, as a consequence of living in a Western, United States context, have fewer choices afforded to them in life, and so it seems, also in death. As an instantiation of limbo, this poem equips Black life and memories with the permission to choose how they will live, and how they will be remembered.

(Re)membering

In the *post* world that we live in, we are being seduced by the puddles we are standing in and led away from the longer wake (Sharpe, 2016) of history and its implication on the present (Sharpe, 2016). I choose to (re)member, as more than just the recall (Durham, 2014) explicated above, but as a process of piecing back together things that have been separated out (Dillard, 2012). To (re)member is to assume that our identities are constitutive of engaged relationship with others in context. Active in nature, it is an "interpretive process of *bridging lived experience with living memories embedded* in words, acts, objects, or sounds to generate temporal, plural, partisan, and partial meaning that is filtered through a historically produced subject" (Durham, 2014, p. 61). In the poem, the intent was to bring my body, living memory, both past and present, together. Remembering comprises the three-step process of interpretation, interaction, and relationality (Durham, 2014; Hall, 1975).

The interpretation stage invited me into a deep, full read of cultural texts to parse out hegemonic sediments and surface the lived experience that the *I* brings to the memories bound up in a text. Here, a full read meant examining media coverage, court proceedings, and testimonies. After performing a full soak of the text, Hall (1975) suggests that the assessment or interaction should consider why the textual bodies are what they are by situating them in their social cultural setting. What do they represent and to whom? In doing so, I wrestled with the possibility of limbo in relationship to the function of power that places the text/ body on the line for exhibit and consumption. The third step involved relationality, and considered how "interaction comprised shared moments and shared meaning between bodies" (Durham, 2014, p. 63). Relationality holds the understanding that texts/bodies constitute each other, complicate each other, and are open to mutual illumination.

Representation

According to hooks (1999), "every woman's confessional narrative has more meaningful power of voice when it is well crafted" (p. 68). I understand this sentiment not as a call for essentialist form; rather, the various manifestations of limbos (James, 1999), are "as much situated in craft as in their epistemological potential" (Spry, 2011, p. 106). Said differently, the depth of knowledge generated and thereby made accessible is directly related to its aesthetic dimension, and in doing so, a sociopolitical action. Limbos (James, 1999), as inherently social, follows hooks' (1999) consideration:

> unless we able to speak and write in different voices, using a variety of styles and forms, allowing the work to change and be changed ... there is no way to converse across borders to speak to and with diverse communities.
>
> *(p. 41)*

Performative expression is the process by which aesthetic choices instruct bodies in motion and space, recognizing that the body is capable of language. Representation in this sense composes the body into being as an intentional, critical, dangerous intervention into the routine orders of social life (Pollock, 1998). The above performance, "Buried," uses poetics to make claims about the tensions and vulnerabilities inherent in Blackened choices to survive. As an embodied writing form, this method of representation is co-performed, that is, it illustrates our conflicting and/or differing experiences of power and privilege; is evocative and encourages sensory engagement, or felt sensing of reactions and interactions; it is consequential, which is to say implicated and not innocent in how it challenges or is complicit in the maintenance of power; it allows the researcher-interpreter-performer "to manipulate language to create sound, to render rhythm, evoke emotion, illustrate ideas, and incite action by using literary devices" (Durham, 2014, p. 103).

Ethics

Representation (Durham, 2014) in this methodological approach needs to be mindful of ethics. For example, appellation of Till's death in this method is not done without consequence. Despite my intent to repurpose the memory as instructive in current and future imaginings of agency, my use of difficult memories (and the general deployment of Black pain and suffering) may contribute to the Black body's general association as the frequent and familiar terrible spectacle (Hartman, 1997). Against this backdrop, "how, if at all, does one give expression to these outrages without exacerbating the indifference to suffering?" (Hartman, 1997, p. 4). Calling for a heightened awareness of Black resonances in the analysis of Black aesthetics, Henderson (2014) implores that we engage in the *miracle of hearing*, as Black writers/performers "make us listen ... to what they hear" (Kovalova, 2016, p. 105). This approach invites the critic/reader to adopt an approach to analyzing text/body that "embraces an oral/aural aesthetic affirming the hearer's/listener's response to the speaker/narrator/storyteller" (p. 106). It is the position of the reader who thus completes the speaker's story upon witnessing it from, and through, a position of deep hearing and empathetic listening.

Relationships of mutual exchange are built on "the dual and simultaneous processes of *address-ability* and *response-ability*" (Oliver, 2001, p. 16). The responsibility of bodies in the wake (Sharpe, 2016) is to "bear critical witness and provide theoretical testimony" (Henderson, 2014, p. 20). When this happens, it becomes possible to reconceive what it means to be a self, a subject, to have subjectivity, [and] to consider one's agency while in liminality. This theme compels me (and all others who would engage similar memories) to ask and answer the questions: Am I a witness to confirm the truth of what happened in the face of a world destroying capacities of pain and the interiority of Black life? Or, am I voyeur fascinated by accounts of terror? After Hartman (1997), what does the exposure of the violated Black body yield? Is this simply an opportunity for self-reflection? What do I do with what I just experienced, exhumed, or viewed? More pointedly, "how does one tell the story of an elusive emancipation and travestied freedom" (Hartman, 1997, p. 10)? Ethical engagement requires a critical examination of both the historical event under investigation and the self, since the very effort to (re)present reveals "the provisionality of the archive as well as the interests that shape it" (Hartman, 1997, p. 10). Stated differently, the representation (Durham, 2014) that follows analysis requires not only a questioning of the dominant narrative, but also the agentic attempt to salvage and remake archival materials for alternative purposes.

Conclusion

Building on McKittrick's (2006) *site and sight of memory*, this Black feminist performance auto/ethnographic approach seeks to re-present the Black subject in a world that has tirelessly labored to dehumanize and erase the possibility of interior lives—to know it only as the failed object of whiteness, and thus, expendable, disposable, and exploitable. This method desires to reconstruct those interior lives with what remains—not the tools of dehumanization—but the remnants—sounds, music, colors, behaviors, and stories. Black feminist performance auto/ethnography as epistemic, material, and imaginative work, through exploration of the real and the possible, can provide a "route to reconstruction" (McKittrick, 2006, p. 32) for endarkened bodies (Dillard, 2000), which have always made meaning while in motion (Danticut, 2015)—ephemeral and palimpsestic.

Poetics in this chapter offered a picture of doing limbo (James, 1999), engendering what Crawley (2013) calls "the forms by which a group of humans engages the difficulties posed for it in existence, including both its possibility, its survival, or its becoming something else altogether [... and] is about possibility in general" (p. 55). Motioning for what hooks (1989) calls a visionary feminist theory, which is "an integration of critical thinking and concrete experience" (p. 39), I *do* limbo (James, 1999) as a performance in "Buried." This Black feminist arrangement represents aesthetic movements that make of constraint and crawlspaces, capaciousness and exodus, always and in all ways responding to the query, *how do we exist in the bodies that we hold in this historical moment?* (Okello, 2018). Upending popular social narratives that consign Black bodies to respectability as evidence of their social worth, Black feminist limbos (James, 1999) make vivid (demand that one see), while in the wake (Sharpe, 2016), a resistance to rational discourse and policies that problematize Black motion.

Notes

1 The use of wake in this title borrows from Christina Sharpe's (2016) text: *In the wake: On blackness and being*.
2 The use of limbo in this title references Joy James' (1999) notion of limbos in her text *Shadowboxing: Representations of black feminist politics*. As done above, the use of wake in this title borrows from Christina Sharpe's (2016) text: *In the wake: On blackness and being*. This Black feminist approach builds on the work of Black feminist autoethnographers such as Anderson, 2008; Boylorn, 2013; Durham, 2014; Hill, 2014; Jones, 1997.

References

Alexander, J. M. (2005). *Pedagogies of crossing: Meditations of feminism, sexual politics, memory and the sacred*. Durham, NC: Duke University Press.

Bailly, J. C., & Anderson, M. H. (2010). The slightest breath (on living). *CR: The New Centennial Review, 10*(3), 1–11.

Baldwin, J. (1985). *The price of the ticket: Collected nonfiction 1948–1985*. New York, NY: St. Martin's Press.

Bell, E. (2008). *Theories of performance*. Thousand Oaks, CA: Sage.

Callier, D. (2018). Still, nobody mean more: Engaging black feminist pedagogies on questions of the citizen and human in anti-Blackqueer times. *Curriculum Inquiry, 48*(1), 16–34.

Crawley, A. (2013). Breathing flesh and the sound of black Pentecostalism. *Theology & Sexuality, 19*(1), 49–60.

Danticut, E. (2015). Black bodies in motion and in pain. *The New Yorker*. Retrieved from: www.newyorker.com/culture/cultural-comment/black-bodies-in-motion-and-in-pain.

Denzin, N. (1994). *On understanding emotion*. San Francisco, CA: Jossey-Bass.

DiAngelo, R. (2011). White fragility. *The International Journal of Critical Pedagogy, 3*(3), 54–70.

Diawara, M. (2011). One world in relation: Édouard Glissant in conversation with Manthia Diawara. *Journal of Contemporary African Art, 2011*(28), 4–19.

Dillard, C. B. (2000). The substance of things hoped for, the evidence of things not seen: Examining an endarkened feminist epistemology in educational research and leadership. *International Journal of Qualitative Studies in Education, 13*(6), 661–681.

Dillard, C. B. (2012). *Learning to (re)member the things we've learned to forget: Endarkened feminisms, spirituality, & the sacred nature of research & teaching.* New York: Peter Lang.

Durham, A. (2014). *Home with hip-hop feminism: Performances in communication and culture.* New York, NY: Peter Lang.

Gabbidon, S. L., Greene, H. T., & Young, V. D. (2002). *African American classics in criminology and criminal justice.* Thousand Oaks, CA: Sage.

Gordon, L. R. (Ed.). (1997). *Existence in Black: An anthology of Black existential philosophy.* New York, NY: Routledge.

Gumbs, A. P. (2014). Nobody mean more: Black feminist pedagogy and solidarity. In P. Chatterjee & S. Maira (Eds.), *The imperial university: Academic repression and scholarly dissent* (pp. 237–259). Minneapolis, MA: University of Minnesota Press.

Hall, S. (1975). Introduction. In A. Smith, E. Immirizi, and T. Blackwell (Eds.), *Paper voices: The popular press and social change, 1935–1965* (pp. 11–24). Totowa, NJ: Rowman and Littlefield.

Hartman, S. V. (1997). *Scenes of subjection: Terror, slavery, and self-making in nineteenth-century America.* Oxford, UK: Oxford University Press.

Henderson, M. G. (2014). *Speaking in tongues and dancing diaspora: Black women writing and performing.* Oxford, UK: Oxford University Press.

Hill, D. (2014). TRANSGRESSNGROOVE: An exploration of Black girlhood, the body, and education. (Doctoral dissertation, University of Illinois at Urbana-Champaign).

hooks, b. (1989). *Talking back: Thinking feminist, thinking black.* Boston, MA: South End Press.

hooks, b. (1990). *Yearning: Race, gender, and cultural politics.* Boston, MA: South End Press.

hooks, b. (1999). *Remembering rapture: The writer at work.* New York, NY: Henry Holt.

James, J. (1999). *Shadowboxing: Representations of black feminist politics.* New York, NY: St. Martin's Press.

Jones, J. L. (1997). "Sista docta": Performance as critique of the academy. *The Drama Review (1988–), 41*(2), 51–67.

Jones, J. L. (2002). Performance ethnography: The role of embodiment in cultural authenticity. *Theatre Topics, 12*(1), 1–15.

Kendi, I. (2017). *Stamped from the beginning: The definitive history of racist ideas in America.* New York: Random House.

Kovalova, K. (Ed.) (2016). *Black feminist literary criticism: Past and present.* New York, NY: Peter Lang.

Magolda, M. B. B. (2009). *Authoring your life: Developing an internal voice to navigate life's challenges.* Stylus Publishing, LLC.

McKittrick, K. (2006). *Demonic grounds: Black women and the cartographies of struggle.* Minneapolis, MA: University of Minnesota Press.

Muhammad, K. G. (2010). *The condemnation of blackness*. Cambridge, MA: Harvard University Press.

Okello, W. K. (2018). From self-authorship to self-definition: Remapping theoretical assumptions through Black feminism. *Journal of College Student Development, 59*(5), 528–544.

Oliver, K. (2001). *Witnessing: Beyond recognition*. Minneapolis and London: University of Minnesota Press.

Pollock, D. (1998). *Performing writing. The ends of performance*. New York, NY University Press.

Raiford, L. (2006). Lynching, visuality, and the un/making of blackness. *Nka: Journal of Contemporary African Art, 20*(1), 22–31.

Robson, D. (2010). *The murder of Emmett Till*. Farmington Hills, MI: Greenhaven.

Scott, J. C. (1990) *Domination and the arts of resistance: Hidden transcripts*. New Haven, CT: Yale University Press.

Sharpe, C. (2016). *In the wake: On blackness and being*. Durham, NC: Duke University Press.

Spillers, H. J. (1987). Mama's baby, papa's maybe: An American grammar book. *diacritics, 17*(2), 65–81.

Spry, T. (2001). Performing autoethnography: An embodied methodological practice. *Qualitative Inquiry 7*(6), 706–732.

Spry, T. (2011). *Body, paper, stage: Writing and performing autoethnography*. Walnut Creek, CA: Left Coast Press.

3

CANVASING THE BODY

A Radical Relationality of Art, Body, and Vibrant Materiality

Tami Spry

Put your body into it. I mean right in it. Right on it. Climb in. Lie on the canvas. Feel the scrape of the dried oils, the scuff of the acrylics. Feel the body that pushed the paint. Feel the body that made it. Feel the body made on it. Now speak from this body diffracted through wood and paper and oil and pigment and skin and blood and water. Feel what happens when these things are theorized as much "force as entity, as much energy as matter" when you lay your body down in art.

So. Let's just get right to it, shall we? Let's just get right to the fact of the matter. The fact of the matter is... The fact of the matter is.... The fact of the matter. Always. Is. And that's what I couldn't seem to let go of, break out of. It's what my grief process could not *get with*, would not reconcile. Things went terribly wrong terribly fast after five months in utero; our son's spirit left before his body was born. And it is the body of a stillborn child that no one wants to talk about. Understandably. But it is the defining subject for the mother. Not talking about it erases the body, making the child, in a very real way, void of force and energy and agency, making a significant part of the grief process invisible and inaccessible.

And so I wrote and storied and reflected and critiqued. I made words and pages and put them on stages, which saved my life. It did. But still something was missing. And then I come across Barad's (2007) question: "How did language come to be more trustworthy than matter? Why are language and culture granted their own agency ... while matter is figured as passive and immutable?" (p. 120). Something in my grief process shifted with these questions. Where was the accountability to the body that he inhabited, even for a short time? His body never having been sentient in this world became a passive object. How might the processing of grief change if his body were considered vibrant matter, a diffraction of matter and meaning and language all with its own agentic properties. A diffraction of my body, of his, of the hospital bed, the time of day, his father's body that I clutched to and hung from before during and after. The matter of the faces of my family and friends as they looked into the room or sat or cried or just read. The matter of the big John Deere lawn tractor that I wouldn't let myself get off of cuz the acreage needed to be mowed before everyone got there for the wake. The lawn. It matters. I mow straight lines. The grass smells alive. Awake. Alive. He's not. The matter of the padding between my legs straddling the tractor, catching the what's left behind the straight lines.

Put your body into it. I mean right in it. Lie on the canvas.

I'm in the Walker Art Center for the national opening of the Frida Kahlo Exhibition. It's 2007. Four years since the fact of those five months. Four years of mowing straight lines. Therapy, writing, and performing, reflecting, critiquing, words words words. I turn the corner and there it is. Kahlo's painting "Henry Ford Hospital." Her child's birth and death. And my jaw drops and my knees buckle, not out of anguish or grief. Jaw drops knees buckle with relief with release. In "The Inertia of Matter and the Generativity of Flesh," Diana Coole (2010) writes that a painting can "break the 'skin of things' and show them emerging into visibility" (p. 105). Finally, as a matter of fact. Finally, someone broke the skin.

And before I know it, I am climbing onto the canvas and into her bed. Where six umbilical cords attach a baby body, a medical drawing of a uterus, a uterus lying in an orchid, a snail, a medical device, and the bone structure of her pelvis. A break in the skin of things.

And our elbows scrape against the canvas as we lay our bodies down at once fractured and mended, back broke, womb torn, separated even from the milk she consumes at the breast of her black-masked nurse maid and all I want to do is a fist pump, a cheer at the diffraction of this life and death experience, at the assemblage of this baby body "as much energy as matter" at once unbearable and vibrantly agentic. A real-izing the agentic matter of this post-present-traumatic-seismic stress disorder displacement of blood and oil and skin and bones and bodily presence and absence and with the paint against my skin the molecules of my human sense of hierarchy forget themselves and I slip and slide through the air with her

> to the bottom
> which is never really the bottom at all;
> it is performative entanglement
> of bodies and words and things
> all the way down
> an enduring plumb into the deep ecology of body-page-body-stage
> inviting what else we might know of being in semantic-somatic multi-
> matter knowing.

Performance embodies the collective agency of things. "A posthuman perform-ance," write Travis Brisini and Jake Simmons (2016), "is … a way of working within the newly heightened constraints of worldly belongings after humanist sovereignty falters" (p. 193).

As Dwight Conquergood reminds us, "For performing researchers the body becomes the porous boundary of exchange, the interface" (quoted in Hamera, 2013, p. 307); the porous exchange of biology, words, and things offered my grief process an accountability to the body that he inhabited, an accountability that could not be satisfied with words alone. The studied doing of performance offers enfleshed potentialities of engaging agentic vibrant matter. Feel the body that pushed the paint. Feel your body in it.

Within the porous boundaries of performance, something vibrates just below the skin, something quivers, judders, trembles under the gut, below the throat, inside the palms. It is affect and more. It is the body and more. It is good art and more. It is an arrangement of atoms forming paint on a surface and more. Viewed through a lens of thing-power (Bennett, 2009), our bodies engage in more than a sensual experience with art. "The body surges," writes Kathleen Stewart (2007); "Agency lodged in the body is literal, immanent, and experi-mental" (p. 113). The performing body radicalizes relations with ink, atoms, oil, affect. Narrative scholar Jane Speedy (2015) began painting after a massive stroke. Her art work explicates the complexities of "agency lodged in the body." "Borders between fact and fiction were interrogated," writes Speedy, "and thresholds between different forms of realism: agential, critical, and/or magical/were breached" (p. 41). Jane paints a posthuman performativity where agential, critical, and/or magical borders of body and thing are breached and blurred. Lie on the canvas.

A radical relationality of the material is what performance is and what it does; not a "demonstration" of relations, but a radicalizing of relations between semantic-somatic things with indiscreet transient borders and being (Spry, 2016). It is Craig Gingrich-Philbrook's (2001) claim, "My body makes language. It makes language like hair" (p. 3). Engaging matter as agentic through Kahlo's work offered me a material belonging with my son that had been missing. Her body and/as brush blurs what we might consider materially "literal." In her *Diary*, Kahlo writes, "we direct ourselves toward ourselves through millions of beings—stones—bird creatures—star beings—microbe beings—sources of our-selves" (p. 24). I needed to know that agency could be located in more than just the living human body, that matter is "an intra-active becoming—not a thing but a doing, a congealing of agency" (Barad, 2007, p. 151), which is of course the very substance of performance, a doing, a thing undone, a thing becoming. "The point," writes Barad, "is not merely that knowledge practices have material consequences but that *practices of knowing are specific material engagements that parti-cipate in (re)configuring the world*" (p. 91, emphasis in original).

So put your body into it. I mean right in it. Speak. From this body.

References

Barad, K. (2007). *Meeting the Universe Halfway*. Durham, NC: Duke University Press.

Bennett, J. (2009). *Vibrant Matter: A Political Ecology of Things*. Durham, NC: Duke University Press.

Brisini, T. & Simmons, J. (2016). Posthuman relations in performance studies, *Text and Performance Quarterly*, 36(4), 191–199.

Coole, D. (2010). The Inertia of Matter and the Generativity of Flesh. In. D. Coole and S. Frost (Eds.) *New Materialisms: Ontology, Agency, and Politics* (pp. 92–115). Durham, NC: Duke University Press.

Gingrich-Philbrook, C. (2001). Bite your tongue: Four songs of body and language. In L. C. Miller and R. J. Pelias (Eds.), *The Green Window: Proceeding of the Giant City Conference on Performative Writing* (pp. 1–7). Carbondale, IL: Southern Illinois University.

Hamera, J. (2013). Response-ability, vulnerability, and other('s) bodies. In E. P. Johnson (Ed.), *Cultural Struggles: Performance, Ethnography, Praxis* (pp. 306–309). Ann Arbor, MI: University of Michigan Press.

Kahlo, F. (1995). *The Diary of Frida Kahlo: An Intimate Self-Portrait*. Ed. C. Fuentes. New York: Abrams.

Speedy, J. (2015). *Staring at the Park: A Poetic Autoethnographic Inquiry*. Walnut Creek, CA: Left Coast Press.

Spry, T. (2016). *Autoethnography and the Other: Unsettling Power Through Utopian Performatives*. New York: Routledge.

Stewart, K. (2007). *Ordinary Affects*. Durham, NC: Duke University Press.

4

INTERSECTIONALITY IN EDUCATION RESEARCH

Methodology as Critical Inquiry and Praxis

Venus E. Evans-Winters and Jennifer Esposito

Our nation's student population is becoming increasingly more diverse and yet many educational practices and policies remain the same. Intersectionality, which began as a Black feminist theory examining the ways multiple oppressions manifest in a person's life (Collins, 1990; Crenshaw, 1991), is a tool that can and should be used to explore the experiences of students and educators in K through post-grad schools. This chapter engages intersectionality through a methodological lens, specifically. We are interested in exploring how intersectionality might be utilized as a conceptual framework in critical qualitative inquiry within education and other closely related fields. We are also interested in its limitations and strengths as both a methodology and a framework for critical methodologies. We ponder the following questions: Has education research taken up intersectionality in all of its complexities? How can the critical framework as methodology and praxis help critical scholars radically excogitate matrixes of domination?

Centering Intersectionality

Intersectionality as a conceptual tool for understanding and interrupting social phenomena evolved from within Black feminism as theory and praxis. As an analytical device, intersectionality examines the ways multiple social–political positionalities manifest and shape individuals' and groups' lives, especially girls' and women's social realities, across socio-cultural and political contexts. From a Black feminist perspective, intersectionality concerns itself with how racism, sexism, classism, heterosexism, and xenophobia, and other interlocking systems of oppression impede the rights and dignity of Black women and other people of color.

Like most social theory, intersectionality did not evolve in a social vacuum nor was it conceptualized by a lone social scientist; however, most agree that the feminist critical race theorist Kimberlé Crenshaw coined the term *intersectionality* in 1991 to describe how Black women and other women of color experience workplace discrimination and violence based on multiple intersecting identities. More specifically, Crenshaw (1991) demonstrated that Black women were excluded from industries that recruited women, because they were not white women; and Black women were not hired by industries that recruited Blacks, because they were not men. Yet, Black women had no case before the courts, because the problem they confronted of hiring discrimination did not affect all women, nor did it affect all Blacks. Crenshaw may also be remembered for her astute analysis of the treatment of Anita Hill, who accused then Supreme Court nominee Clarence Thomas of sexual harassment.

It is important to note that women-of-color feminists throughout history have been concerned with the ways multiple oppressions shape lives. When Sojourner Truth asked "Ain't I a Woman" (see Stowe, 1863) at the Seneca Falls Convention on Women's Rights, she was articulating how difficult it was to be both a woman and a Black person, as neither had true freedoms. Historically in the U.S., and in contemporary times, Black women, as well as other women of color, carry the burden of both racial discrimination and gender discrimination and other kinds of discrimination. We see intersectional writings and speeches throughout many civil rights movements with women of color including Latinas, Chicanas, and indigenous women, writing anthologies about intersectional concerns. Most notable is Cherrie Moraga and Gloria Anzaldúa's *This Bridge Called my Back* first published in 1981, as well as Anzaldúa's *Making Face/ Making Soul: Haciendo Caras* originally published in 1990.

Women of color's multiple marginalized identities place them in jeopardy of intersectional forms of oppression asserts Crenshaw (1991). Accordingly, intersectional theorists consider how people are multiply situated and how coercive power and systematic oppression cannot be fully understood by asynchronous examinations of structural or relational power. Intersectionality recognizes that identities are mutually interlocking as well as relational (Berger & Guidroz, 2009; Collins, 1998). Prior conceptions of social identity imagined individual and group identities as additive and ordinal, with one identity being the primary while other identities were subsequent or secondary to the main identity.

Consequently, assumptions about oppression were also imagined to be additive or ordinal as opposed to interlocking, dynamic, and complex. As a theoretical framework, intersectionality is not primarily about identity, but about being conscientious about the means by which social structures make certain identities vulnerable. As an analytical tool in qualitative inquiry, intersectionality requires us to ponder our interpretations of individual and social identities, human relationships, and social environments in more complex ways. How can intersectionality as a framework in qualitative inquiry inform anti-racism in education and feminist politics in education? How can intersectionality in education research push for examinations of patriarchy in anti-racist inquiry and simultaneously push against racial hierarchy in feminist qualitative inquiry?

Authors' Positionality

I, Venus, identify as a Black woman scholar and mother. My scholarly and personal identities are influenced by (the tensions between) Black feminism, critical race theory, Afrocentrism and Pan-Africanism. I spent my earlier years attending schools on the southside of Chicago, and the later adolescent years in a poor and working-class suburb of Chicago. Growing up in the midst of the war on drugs (or poor Black communities), I align aesthetically with hip hop culture and politically with Black liberation struggles. It is my personal belief that formal education is a vehicle for social change and a tool of domination. Similarly, I believe that Western science has not always served the interest of my people and the various communities that I represent. Suffice it to say, I am interested in engaging efforts to decolonize research methodologies and interrupt epistemic apartheid. Zora Neale Hurston is my methodological role model.

I, Jennifer, have worked (through research and writing) alongside Venus since graduate school. From our initial conversations about researching our people, we knew that we were soul sistas and kindred spirits. I identify as Latina, specifically Cuban, though I was raised in mostly Black neighborhoods in New York. My ancestors' pain and trauma is alive in my DNA and provides guidance and a road map when necessary. I have always been interested in the spiritual and the Orishas. My interest in poetry at a young age led me to Audre Lorde and Cherrie Moraga. Their words lit my soul on fire and I could feel my ancestors with me as I read. I am a former K-12 teacher, a feminist, and methodologist. My scholarship has always tried to examine the ways multiple marginalities shape people's access to and experiences within education. The creative desire to use words to express pain and fight against injustice has never left me and, thus, I engage in decolonial research methodologies as much as possible.

Intersectionality in Educational Research

Arguably, intersectionality in qualitative inquiry would require that we examine our own identities to discover how they play out in the research process, in addition to grappling with how interlocking systems of oppression like white supremacy, patriarchy, heterosexism, capitalism, and imperialism impede the everyday lives of those whom we study and the cultural context where the study takes place (e.g., schools, institutions of higher education, urban or suburban neighborhoods, workplaces, etc.). In recognition of our foremothers and feminist critical race theorists, framing critical qualitative inquiry from an intersectional perspective, our starting points are efforts to de-marginalize the intersection of race and gender (Crenshaw, 1991, 1989) in scholarship.

An assumption of those of us who embrace intersectionality as (embodied) theory and praxis is that our in/visible identities, relationships, and interpretations of our experiences are complex; therefore, our approaches to inquiry and analysis will also be complex, messy, and multilayered. As Collins and Bilge (2016) explain, "Intersectionality is a way of understanding and analyzing the complexity in the world, in people, and in human experiences.... Intersectionality as an analytic tool gives people better access to the complexity of the world and of themselves" (p. 2). Purportedly, intersectionality begins to move us away from so-called identity politics and claims of essentialism reminiscent of feminism's past.

Power, Privilege, and Domination

As an analytic framework, intersectionality posits that although institutional structures, social norms, and authority may appear static and resilient in the face of social change, the proximity of individuals and groups to power may shift over time or across contexts. Conversely, intersectional frameworks also proclaim that certain groups of people have historically been systemically cemented to the bottom of society, exhibited resilience and resistance in the face of oppression, and challenged social norms that brought about meaningful (albeit slow and sometimes quelled) social change. In the U.S. context, for instance, enslaved Africans and Black Americans, and indigenous and poor people have historically endured systematic de jure and de facto judicial, economic, housing, and educational disenfranchisement.

We argue here that critical qualitative inquiry from an intersectional perspective must explore: (1) how power and authority are simultaneously fixed *and* astatic within and across social contexts; (2) how individuals or social groups resist, confront *and/or* placate oppressive authority and structures of power; (3) how space (social and spatio-temporal) influences how one perceives and enacts power; and, (4) how one's "place" (as an individual and as part of a social group or groups) in history and contemporary society shapes one's approaches to qualitative inquiry and research knowledge productions. Inevitably, critical qualitative inquiry from an intersectional perspective unapologetically and enthusiastically acknowledges that people can be multiply situated in the world.

As a standpoint theory (Collins, 1990; Smith, 1987), qualitative researchers who embrace intersectionality as an analytical tool believe that the multiple perspectives of the marginalized and oppressed offer unique and often divergent viewpoints of the social world. Some feminist critical race scholars, like us, also seek to draw upon our own cultural knowledge and intuition (Ahmed, 2017; Delgado Bernal, 1998) to counter hegemonic forces and universal claims of truth that have caused undue harm to our people through misrepresentation, medical exploitation in the name of science (see Washington, 2006), social control and management, and militarization and colonization. It is through a shared understanding of the social world that women of color, the indigenous, racial and ethnic minoritized people, queer and non-gender conforming people, and the poor, especially, can draw upon intersectionality as an analytical tool in the examination of (and resistance to) systems of power, privilege, and domination.

Whose Ways of Knowing?

Recently I attended a national research conference that attracts education researchers from all over the world. I've been attending the conference annually for more than 15 years, since I was a graduate student. At the most recent annual conference, I was invited by a graduate student to attend her paper session. As it was explained to me in a state of jubilation, the graduate student used my past work in her dissertation, and she was proud but nervous that her paper had been accepted for presentation. She was excited that her research paper would be presented at a major research conference, and she would appreciate having one of her mentors present in the room. When I arrived approximately five minutes late to the paper session, I made my way up to the front of the room to sit next to a prominent scholar in the field of multicultural education.

The session attracted a racially diverse crowd with graduate students of color visibly representing to support their own. In fact, the panel comprised mostly graduate students, and all but one of the presenters were Black. A few papers discussed research on Black girl's and women's schooling or experiences in educational settings. When the white woman began to speak, she described herself as "queer" and mentioned that she utilized "intersectionality" as a theoretical lens, because her research participants were "Black urban students" who were also poor and identified as having disabilities. She then explained that intersectionality was the best framework, because she too had multiple identities.

Let's just say that I wasn't turned on or off by her research discussion. However, it was more than obvious that others around me were turned off by the presentation. For example, the senior scholar next to me clicked her teeth and stared me up and down every time the presenter mentioned intersectionality. Other Black scholars looked at me to witness how I was receiving the presentation and claims made by the white presenter. Somehow in that moment, I became the resident Black feminist expert. In that moment, I refused to validate or discredit the speaker's claims, and also purposely mentally noted the audience's reaction. By the time the chair announced it was time for questions, audience members either actively ignored the white queer presenter or directly questioned her claims of intersectionality.

For hours and now months following that uncomfortable exchange in the midst of a major educational research conference session, I (Venus) thought extensively about the evolution of intersectionality and its utility to educational research, policy, and practice.

First, we must acknowledge that the majority of researchers present in the front of the conference room that day were actually young Black scholars who chose to examine their own marginalized experiences in education spaces, or they conducted research with Black students in schools. Subsequently, we also recognize that many of those present, in the audience and on the panel, questioned the motives of a white woman, although queer identifying, researching Black youth. Third, there was apparent concern about who would legitimately become an authority of Black urban education.

Personally, we believe it is redundant and goes beyond the scope of this chapter to discuss the question raised repeatedly in qualitative research about who gets to research whom. However, we believe it is important from a critical qualitative perspective to examine how education research is received and perceived by both the educational researcher and audiences of education research, using this particular case. In other words, before we can contemplate the adoption of intersectionality by the researcher, we also must acknowledge the research audience. The face of public education and higher education in the U.S. is changing. Black people and other people of color are more and more becoming the producers and consumers of knowledge. Thus, our knowledge pursuits and criteria of trustworthiness and authenticity in qualitative inquiry will necessarily shift across cultural communities and contexts.

In the particular case described above, graduate students of color came prepared to support fellow graduate students in a seemingly hostile or unwelcoming space. There is obviously emotionality in pursuits of truth and in truth claims. They understood that different people are differently rewarded (and punished) in academia, and many of them know firsthand how scientific research has exploited, dehumanized, and mischaracterized people who look like them and come from where many of them come from. Consequently, as the faces of academia change, we can expect to witness not only a researcher's interpretations scrutinized, but also a scrutiny of *the motives behind* the researcher's choice of education topic, research participants, research context, research methodologies, cultural understanding, etc.

Historically, science and truth claims in the peer review process were depicted as objective, value-free, and for the good of society. Similarly, qualitative researchers admitted that the scientific method was inherently subjective and value-laden. But, even at times, qualitative research itself fails to understand and expose how researchers and the research process itself can simultaneously interrupt and perpetuate cultural hegemony. Below, Jennifer contemplates how her own identities and cultural affiliations facilitated compromises, negotiations, and pushed the boundaries:

As articulated in my positionality statement, I (Jennifer) have collaborated with Venus for many years. When I look at my body of scholarship, I see more focus on Black children and people than I do to my own Latinx community. Part of this is who I aligned myself with growing up, who embraced me given that I was noticeably out of place. Part of this is who I read from an early age that resonated with me. I have been an intersectional feminist my entire academic career.

From my earliest publications almost 20 years ago, I cited Black and Brown women. Their words, like Lorde's and Moraga's, lit my soul on fire and continue to do so. I would not want the boundaries of who could research whom policed in such a way that my work for and with Black women became invalidated solely because I am not a Black woman. But I know that my work may be subject to a different type of criticism or that I may have to build a different type of trust with my readers than Venus might. I am okay with this. I am not the authority on Black (or Brown) women. Nor do I want to be. But I do want to continue to do work that pushes boundaries and recognizes that Black and Brown people, though marginalized in many ways, have always resisted and survived. I am made in my Abuela's image.

My issue with some intersectional research is that it has become a code for the combination of race and gender. It is a fad now to do intersectional work. The sad truth Is, though, that the term has been co-opted. And, once something is co-opted, it loses its power. I don't want to read another study that included Black girls, and that calls itself intersectional, but that does not do the hard work of interrogating structures of power and how our society's investment in these structures of power perpetuates oppression.

Critical Reflexivity

Therefore, from an intersectional perspective in qualitative inquiry, instead of asking who has the right to research whom, we might ask from an intersectional perspective, what is the researcher's social political proximity to the educational problem or issue, and to the research participants themselves? We must be cautious of any scholar's claims that their selected theoretical framework offer them "insight" into marginalized groups' experiences. Black feminism, queer theory, Marxism, intersectionality, or any other critical theory alone does not give one some type of exclusive awareness of others' ways of life or spontaneously foster empathy or mutual suffering.

An intersectional approach in education research requires conscience reflection and thoughtful awareness of how multiple and interlocking oppressions impact similarly or differently the researcher and research participants. Critical reflexivity fosters opportunities for relationship building and advocacy before, during, and after the research process. Intersectional approaches in qualitative inquiry with a clear critique of power also unabashedly attempt to adapt methodologies that serve to help educe the power (Akbar, 1999) of research participants, center their multiple ways of knowing/multiple consciousness (Dill, 1979; hooks, 1981; Collins, 1986, 1990; King, 1988), and expose the inherent bias and often unequal rewards of the education research process.

Finally, in the research example above, the presenter was openly challenged for what an audience member interpreted as the presenter's equating being "a lesbian with being Black." I (Venus) recall feeling homophobia played a role in rejecting the presenter's discussion and presence in the group. Would the presenter have been received differently if her research participants were non-gender conforming or LGBTQ students or simply identified as all white students? Can having a queer experience in a homophobic society equate to a genuine concern and yearning to understand the vulnerability of Black youth? We cannot answer those questions here; however, we can acknowledge the fear of and patterns of epistemic apartheid in academia generally and feminist scholarship specifically.

Oddly enough, I do recall being irritated during the research presentation described above, because the presenter mostly cited white feminist theorists while proclaiming to engage the research from an intersectional perspective. Even though she cited the Black feminist lesbian poet Audre Lorde in her presentation a few times, the presenter rarely or never mentioned to my knowledge the intellectual foremothers of intersectionality, including Sojourner Truth, the 1979 Combahee River Collective (republished in 2014), or Kimberlé Crenshaw. This oversight is problematic because Black women intellectual activists are erased from feminist history and intellectual history.

Historically, white women have appropriated Black women's knowledge claims without giving much credit to Black women social and intellectual thought leaders. The most memorable moment of co-optation and overshadowing of Black women's political struggles was the women's suffrage movement of the late 1800s and early 1900s. While Black women supported and organized for women's and Black people's right to vote, white women turned their backs on the Black woman. Using the organizing and advocacy *methodologies* of abolitionists, White women advocated for the suppression of the Black vote in support of white women's suffrage, but not all women's suffrage—just white women's right to vote (Davis, 1983). This early history and suppression of Black women's civic and political pursuits play out today.

As an example of the latent consequences of erasing Black women's history and thought, we present below a pedagogical reflection from Venus's Black Feminist Thought class.

> We started our semester with course readings and discussions on the organization of Black family living on southern slave plantations and the division of labor amongst enslaved women and men. Next, we discussed various civil rights movements and Black women's roles in the movements. As a part of this discussion, we discussed relevant policies that derived from these movements like women's suffrage, *Brown vs. Board of Education*, and then the Civil Rights Act of 1964. Our class discussions were mainly circular as opposed to linear.

Nevertheless, it was on the day of the discussion about the role that Black women played in the 1960s Civil Rights' movement when a Black woman student raised her hand to ask a question. The expression on her face was perplexed, and I could see that she did not know how to phrase her question.

After I called on her by name she exclaimed, "I am confused. I thought that women gained the right to vote in the 1920s or something."

I said, "Yes, but…"

The whole class became quiet and some exhibited similar physical signs of confusion as their classmate who raised the question.

The student continued, "So, now you saying that the Civil Rights Act allowed Black people to vote. Wait. Black women couldn't vote until the 1960s?"

"Yes."

"That's mind blowing. I'm a history education major and all my life, I've been taught that women were allowed to vote in 1920. And, that's what I write on my tests. I never thought about the fact that Black women did not have the right to vote!" The student appeared disappointed in herself and in her schooling.

The example above demonstrates a distortion of the historical record and how it can cause actual cognitive dissonance or distrust among students of feminist-based knowledge. Another example of young women and men watching before their own eyes Black women's intellectual and cultural labor be co-opted is the current "#MeToo" movement. Tarana Burke coined the term "Me Too"—in 2006—to bring attention to the effects of sexual violence and harassment on the socio-emotional health of poor women of color (see Olheiser, 2017). Burke's message attempted to erase the shame of sexual harassment and assault by letting survivors of sexual violence know that they were not alone in their pain and had empathetic supporters in their quest to seek help.

Interestingly enough, one day, we all turned on our social media feeds and televisions and discovered that a white woman captured the phrase, at first with no credit to Tarana Burke, and the phrase quickly went viral (e.g., #MeToo). The phrase "Me Too" and white women became synonymous with workplace sexual harassment and violence. Wealthy white women became the face of the pervasiveness of abuse against women in the workplace, and women of color and poor women became erased from the national conversation on sexual assault and violence. Consequently, media and public attention, resource allocation, and policy initiatives will be directed away from the most vulnerable in society.

In sum, scholars of color understand the political, economic, and social impact of being erased from the historical and scholarly record. We understand that some people's knowledge claims will be taken more seriously and viewed as more objective than others' assertions and formulations. Intersectionality in qualitative education research moves us beyond simplistic notions of identity and toward an accepted understanding that our statuses differently position us in society and in the research process. Of course, questions of credibility and authenticity across differences (i.e., age, race, social class, abilities, language, gender, sexuality, social status, etc.) can be resolved through more collaborative and symmetrical approaches to qualitative research, such as youth participatory action research.

Framing Intersectional Methodological Invocations

Collins and Bilge (2016) proclaim that the core ideas of intersectionality are social inequality, power, relationality, social context, complexity, and social justice. Therefore, in our contemplations of intersectionality in education as critical qualitative inquiry, we must consider not only how and at what *entry point* scholars theoretically engage with/in intersectionality in the research process, but more importantly how in our methodological invocations education researchers engage in conversations on:

- The ways in which social inequality persists through the scientific research process itself and in schooling;
- How power (and proximity to power) influences one's decision to conduct research, research questions, methods, choice of participants and research contexts;
- How power influences relationships in education research and/or education contexts;
- Ethical considerations for building coalitions and more symmetrical relationships with research participants, communities, and across/within organizations to mediate or combat oppressive schooling;
- How education research sometimes unveils and at other times obscures the complexity of social identities, social statuses, and human relationships inside and outside of schools;
- The utility of education research for supporting communities' social justice efforts and addressing issues of equity in school communities, relationships, curriculum, and/or policy.

With the above questions in mind, intersectionality in education research reveals the limitations and the possibilities of science for the most vulnerable and marginalized in society. For instance, in a review of research in the field of education, Hernández-Saca, Gutmann Kahn, and Cannon (2018) used a socio-historical approach to examine how youth and young adults with disabilities are mutually situated in terms of race, gender, social class, sexual orientation, and other social identities.

Interweaving the theoretical tenets of critical special education studies, dis/ability studies, and intersectionality, the authors analyzed research reports "to better understand the political, emotional, socio-cultural, and historical contexts of disability and its effects on lived experiences in schools" (Hernández-Saca, Gutmann Kahn, & Cannon, 2018, p. 287). It is noteworthy to mention that the researchers began their discussion by presenting their own positionality. The three collaborating authors individually and together represented multiple social locations within and across racial, ethnic, linguistic, religious, sexual, geographical, professional, familial, etc. landscapes. Drawing from their own individual and collective interpretive lenses and a social historical approach, the researchers concluded that youth participants depicted in the studies reviewed created intersectional "feeling—making" meaning (p. 294) of their dis/abilities.

The authors' methodologies and conclusions offer various implications for intersectionality in qualitative research. In particular, we are reminded of the significance of the researcher's positionality and how it humanizes the research process and reveals the researcher's potential analytical and subjective insights into the research topic or issue at hand. Second, the study highlights the significance of the exploration of the relationship of power and discourse, and how it plays out in students' everyday lives, such as the label "disability" traditionally having negative connotations, especially for poor, Black and Brown people, and girls.

In other works, Venus discusses specifically how Black girls, in particular, fall through the cracks in special education discourse (Evans-Winters, 2016) and gifted education research and policy (Evans-Winters, 2014), because of the lack of intersectional perspectives in special education theory. Black people, especially poor Black and working-class families, are not supposed to be gifted, and a history of segregating and misdiagnosing Black students' behavior and intellectual capacity has left many Black parents skeptical of the special education process. Notwithstanding, Hernández-Saca, Gutmann Kahn, and Cannon (2018) mention in their analysis the historical association of educational disabilities with Eugenics science, notions of white supremacy, and ableism.

Even so, Hernández-Saca, Gutmann Kahn, and Cannon's research analysis not only focuses on the vulnerability of youth due to their intersectional marginal identities, but also humanizes the youth and young adults in the studies by emphasizing how participants connected their socially constructed identities with their emotions to make meaning of their lived experiences. Thus, youth were not simply passive objects of study for interpretation, but they were active participants in their own lives.

In closing, using this one study cited above as an example alongside our own lived experiences, we attempt to illustrate how intersectionality in educational research strives to synchronously humanize the researcher and researched, emphasize the vulnerability and agency of participants and researchers, and show the complexities of the relationship between power/authority and people's meaning-making of their identities and those social conditions that shape their identities. All the aforementioned functions to move critical qualitative inquiry toward more provocative cultural critique and radical action.

Conclusion

In sum, we asseverate that recently there has been an increase in scholars across disciplines who purport to use intersectionality theoretically and to draw upon intersectionality methodologically. Moreover, with the seeming increase in student "diversity" in pre-K–12 and college/university campuses, and the institutionalization of diversity initiatives over the last decade (Ahmed, 2012) in higher education, it is worth exploring the benefits of intersectionality for teaching, learning, and curriculum. Furthermore, we suggest that with a reinvigorated intellectual interest in critical theory (e.g., Black feminism, queer theory, critical race theory, decolonization theory, etc.) across disciplines, possibly due to the visibility (i.e., mainstream and social media coverage) of grassroots organizing against state sanctioned violence against Black people (e.g., the Say Her Name campaign, #BlackLivesMatter), sexual harassment in the workplace (e.g., #MeToo), sexual violence as the norm (i.e., the Kavanaugh hearing and allegations against number 45 in the White House), and the erosion of basic human rights and civil rights for immigrants, it is more important than ever to examine the role of educational research in unveiling, interrupting, and mitigating injustices in schools and/or providing strategies to combat social injustice and all the ways it manifests across social contexts.

What is a shared modus operandi for comprehending education and schooling from a social justice perspective? How has educational research taken up intersectionality in all of its complexities? We are interested in the limitations and strengths of intersectionality as a methodology informed by a set of concepts, beliefs, and principles (i.e., theoretical framework). Intersectionality is the new "sexy" in education research discourse. Educational scholars may embrace the utility of intersectionality for education theory and practice; however, at this moment, not enough is known about *how* intersectionality functions methodologically in educational research to interrupt educational inequity.

References

Ahmed, S. (2017). *Living a feminist life*. Durham, NC: Duke University Press.
Ahmed, S. (2012). *On being included: Racism and diversity in institutional life*. Durham, NC: Duke University Press.
Akbar, N. (1999). *Know thyself*. Tallahassee, FL: Mind Productions.
Anzaldúa, G. (1990). *Making faces, making soul/haciendo caras: Creative and critical perspectives of feminists of color*. San Francisco, CA: Aunt Lute Books.

Berger, M. T. & Guidroz, K. (2009). Introduction. In M. T. Berger and K. Guidroz (Eds.), *The intersectional approach: Transforming the academy through race, class, and gender*, pp. 1–24. Chapel Hill, NC: University of North Carolina Press.

Collins, P. H. (1998). It's all in the family: Intersections of gender, race, and nation. *Hypatia, 13*(3), 62–82.

Collins, P. H. (1990). *Black feminist thought: Knowledge, consciousness, and the politics of empowerment.* New York: Routledge.

Collins, P. H. (1986). Learning from the outsider within: The sociological significance of Black feminist thought. *Social Problems, 33*(6), pp. 14–32.

Collins, P. H. & Bilge, S. (2016). *Intersectionality: Key concepts.* Malden, MA: Polity.

Combahee River Collective. (2014). A black feminist statement. *Women's Studies Quarterly*, 271–280.

Crenshaw, K. (1991). Mapping the margins: Intersectionality, identity politics, and violence against women of color. *Stanford Law Review*, 43, 1241–1299.

Crenshaw, K. (1989). Demarginalizing the intersection of race and sex: A black feminist critique of antidiscrimination doctrine, feminist theory and antiracist politics. *University of Chicago Legal Forum*, 139.

Davis, A. Y. (1983). *Women, race, and class.* New York: Vintage.

Delgado Bernal, D. (1998). Using a Chicana feminist epistemology in educational research. *Harvard Educational Review, 68*(4), 555–583.

Dill, B. T. (1979). The dialectics of black womanhood. *Signs: Journal of Women in Culture and Society, 4*(3), 543–555.

Evans-Winters, V. E. (2016). Schooling at the liminal: Black girls and special education. *The Wisconsin English Journal, 58*(2), 140–153.

Evans-Winters, V. E. (2014). Are Black girls not gifted? Race, gender, and resilience. *Interdisciplinary Journal of Teaching and Learning, 4*(1), 22–30.

Hernández-Saca, D. I., Gutmann Kahn, L., & Cannon, M. A. (2018). Intersectionality Dis/ability Research: How dis/ability research in education engages intersectionality to uncover the multidimensional construction of dis/abled experiences. *Review of Research in Education, 42*(1), 286–311.

hooks, b. (1981). *Ain't I a Woman. Black women and feminism.* Boston, MA: South End Press.

King, D. K. (1988). Multiple jeopardy, multiple consciousness: The context of a Black feminist ideology. *Signs: Journal of Women in Culture and Society, 14*(1), 42–72.

Moraga, C., & Anzaldúa, G. (Eds.). (1981). *This bridge called my back: Writings by radical women of color.* London: Persephone Books.

Olheiser, A. (2017). The woman behind 'Me Too' knew the power of the phrase when she created it—10 years ago. Retrieved from www.washingtonpost.com/news/the-intersect/wp/2017/10/19/the-woman-behind-me-too-knew-the-power-of-the-phrase-when-she-created-it-10-years-ago.

Smith, D. E. (1987). *The everyday world as problematic.* Boston, MA: Northeastern University Press.

Stowe, H. B. (1863). Sojourner Truth. *Atlantic Monthly, 473*, 481.

Washington, H. A. (2006). *Medical apartheid: The dark history of medical experimentation on Black Americans from colonial times to the present.* New York: Doubleday Books.

PART II
Methodological Inflections

5

VOICE IN THE AGENTIC ASSEMBLAGE

Lisa A. Mazzei and Alecia Youngblood Jackson

In this chapter,[1] we think with Bennett's (2010b) concept of the *agentic assemblage* to position the discursive artifacts of qualitative inquiry as always already material and the material as an always already discursive construction. That is, we move toward positing voice in qualitative educational research as a thing that is entangled with other *things* in an assemblage (Deleuze & Guattari, 1987) that acts with an agential force (Bennett, 2010b). We do this by (re)configuring voice: we refuse the primacy of voice as simply spoken words emanating from a conscious subject and instead place voice within the material and discursive knots and intensities of the assemblage. Thus, we do not "calibrate" voice to the human, nor do we attend to voice as "either pure cause or pure effect" (Barad, 2007, p. 136) via human intentionality. Rather, we account for voice as a material-discursive practice that is inseparable from all elements (human and non-human) in an assemblage.

Like MacLure (2013), we "argue for research practices that would be capable of engaging the materiality of language [voice] itself—its material force and its entanglements in bodies and matter" (p. 658). To account for these entanglements, we follow Barad's (2007) assertion of posthumanism as critically engaging the "limits of humanism" (p. 428). By this, we mean that conventional approaches to educational inquiry often privilege the voice of the humanist subject, assuming that voice can speak the truth of consciousness and experience. In such paradigms, voice lingers close to the true and the real, and because of this proximity, has become seen almost as a mirror of the soul, the essence of the self. Educational researchers have been trained to afford authority to voice, to "free" the authentic voice from whatever restrains it from coming into being, from relating the truth about the conscious self (Jackson & Mazzei, 2009). In this drive to make voices heard and understood, qualitative methodologists have taken up various practices in attempts to "let voices speak for themselves," to "give voice," or to "make voices heard" in educational research (see Jackson, 2003, for an epistemological perspective and critique).

Voice as present, stable, authentic, and self-reflective: such a voice is imbued with humanist properties and thus attached to an individual (be that individual theorized as coherent and stable or fragmented and becoming). Voice is still 'there' to search for, retrieve, and liberate. A refusal to seek methodological practices that attempt to retrieve and liberate voice requires a decentering of the humanist voice that "knows who she is, says what she means and means what she says" (MacLure, 2009, p. 104). An ontological, posthuman consideration of voice moves us from these humanistic practices to emphasize voice as (re)configured in the intra-actions between the material and discursive, as merely one part—and perhaps the least vital component—of an agentic assemblage. As Barad (2007) reminds us, "Posthumanism doesn't presume the separateness of any-'thing,' let alone the alleged spatial, onto-logical, and epistemological distinction that sets humans apart" (p. 136).

We want to emphasize that we are working within and against a particular methodological 'centrism' that privileges both speaking and hearing human subjects; we also attempt to push against the cutting off of a human subject (be it a humanist "reflecting" one or a poststructuralist "contradictory" one) as the prime source of experience, knowledge, and the real (see Snaza & Weaver, 2014). This ontological separation that has permeated educational inquiry has certainly resulted in creating a boundary between 'what has a voice' and what doesn't, thus our interest works against this separation to view a posthuman voice not as a possession but as a thing entangled with other things. To experiment methodologically, we rely on the philosophical frameworks and concepts from posthumanism to disrupt normative and normalizing assumptions of voice in educational research. To do so, we take what is typically called an "interview data excerpt" in qualitative inquiry and rethink it in such a way that the speaking "subject," or the "subject undone" (St. Pierre, 2004) becomes merely one agent in the assemblage. That is, a "subject undone" is not a pre-existent self-contained individual but, instead, is a never fully constituted unfolding produced in social, environmental, technological, and cultural assemblages (St. Pierre, 2004, pp. 288–291).

We will take up a specific example of an entangled voice later in this chapter, but for now, we turn to the broader ontological, posthuman concepts of assemblage and agency. We begin with Deleuze and Guattari's assemblage and move into Bennett's work on the agentic assemblage and the Baradian concepts of intra-action and distributed agency.

Assemblage and Agency in Posthumanism: Bennett's *Agentic Assemblage*

Deleuze and Guattari (1987) state that an assemblage

> comprises two segments: one of content, the other of expression. On the one hand it is a *machinic assemblage* of bodies, of actions and passions, an intermingling of bodies reacting to one another; on the other hand, it is a *collective assemblage* of enunciation, of acts and statements, of incorporeal transformations attributed to bodies.
>
> *(p. 88, emphasis in original)*

Deleuze and Guattari's assemblage is a useful figuration for imagining a posthuman voice: voice is one part of an assemblage that includes multiple, heterogeneous elements. Objects, discursive signs, utterances, bodies—all exist on different temporal and spatial scales that work collectively to produce a territory. The assemblage offers us traction for our reconfiguration of voice because, as Deleuze and Guattari (1987) explain, words and utterances in an assemblage do not have correspondence (or representational resemblance) to anything *but the whole*; they write, "There is a primacy of the collective assemblage of enunciation over language and words" (p. 90). We will subsequently return to this point, but for now, we want to emphasize that everything in the assemblage works as an aggregate: the interminglings produce affects, potentialities, desires (see Ringrose, 2011). The *assemblage* is that which creates a territory and the potential for de-territorialization.

We turn to Bennett (2010a), who writes, "I experiment with narrating events … in a way that presents nonhuman materialities … as themselves bona fide agents rather than as instrumentalities, techniques of power, recalcitrant objects, or social constructs" (p. 47). These nonhuman materialities are bona fide agents that Bennett bestows with what she terms "thing-power" (Bennett, 2010b, p. 2). In what she names an "agentic assemblage," these "things" act with a force. It is not just that "things" in the assemblage act with a force, but the assemblage itself acts, blocks flows, makes cuts, and produces intensities in a theory of distributed agency:

> Assemblages are ad hoc groupings of diverse elements, of vibrant materials of all sorts. Assemblages are living, throbbing confederations that are able to function despite the persistent presence of energies that confound them from within. They have uneven topographies, because some of the points at which the various affects and bodies cross paths are more heavily trafficked than others, and so power is not distributed equally across its surface.
>
> *(Bennett, 2010b, pp. 23–24)*

According to Bennett, "A lot happens to the concept of agency once non-human things are figured less as social constructions and more as actors, and once humans themselves are assessed not as autonoms but as vital materialities" (p. 21). Being faithful to the "distributive quality of 'agency'" (p. 21) is noting that these things, these vital materialities, together in an agentic assemblage, possess agency, not in and of themselves, but in this assemblage, they become an-other body or agent. Thinking voice according to Bennett's framework and her concept of vibrant matter requires an acknowledgment of "things" as being agential, as functioning as catalysts in producing events. Therefore, we no longer think voice as a discrete representation of experience spoken forth by an individual subject, but as an-other body or agent in the agentic assemblage that acts and confounds from within.

If "agency is everywhere" (Hekman, 2010, p. 123), and by that we mean that *agents* are everywhere in the artifacts of our research, in the materiality of our educational field sites, in our analyses, and in our knowledge producing practices, then what is to happen to our practices, our researcher selves, and our thinking qualitative data and data analysis differently if we are to think voice as vibrant (Bennett, 2010b), as agentic, as intersectionally linked with/in forces that we refer to as a posthuman voice? To think voice as vibrant and agentic is not to treat voices as something to be mined in the textual artifacts of educational research, nor is it to ascribe meaning by a focus on what people say, but it is to think voice as that which is produced in the intra-action of things—bodies, words, histories—that, as an assemblage, act with a force. As we continue in this chapter, we will make the argument that it is the assemblage that acts; it is the entanglement that pressures and produces reconfigurings; it is the intra-action of the material and discursive that produces a distributed agency between and among human and nonhuman entities. In doing so, we offer a methodological move to enable a more pro-vocative outcome, a practice that not only pushes against (over-)simplified treatments of voice-as-data but also the humanistic analytic strategies that tend to take for granted what voice "is" and that seek to fix meaning based on one's articulation of "experience." As we mentioned earlier, we intend that this methodological experiment will contribute to the practice of theory and research in education.

Voice in the Kitchen Assemblage

For Bennett (2010b), as for Deleuze and Guattari (1987), bodies and actors in a network, or assemblage, can no longer be thought as subjects and objects. Nor can we any longer think of doers (agents) behind deeds or actions giving "voice" to an experience. There is no hierarchy of primary and secondary agents, nor a deciphering of purpose or intent. Bennett describes how Coole (Bennett, 2010b) "replaces the discrete agent and its 'residual individualism' with a 'spectrum' of 'agentic capacities' housed sometimes in individual persons, sometimes in human physiological processes ... and some-times in human social structures" (p. 30). Coole's notion of the spectrum does not imply development from one end to the other. Rather, Coole proposes the idea of a spectrum to suggest the emergence of agency at different levels of (co-)existence (Coole, 2005, p. 128). She further explains that central to her argument is an insistence that these agentic capacities or properties are "only contingently, not ontologic-ally, identified with individual agents" (pp. 124–125). In the same way that agentic properties are only contingently identified with individual agents, we imagine voice to be so.

Next, we present a research artifact to demonstrate how we conceive post-human voice as not attributable solely to a humanist subject. Rather, we present posthuman voice as a production of a movement of forces in an assemblage. "Things" with agentic capacities come together with other 'things' in an agentic assemblage to produce this posthuman voice. We position this research artifact as an assemblage of heterogeneous elements—objects, signs, utterances, bodies— existing on different temporal and spatial scales: a kitchen table and chairs, gen-dered bodies, discourses of patriarchy and institutionalized religion, enunciations of critique, a garage door, histories, communities, childhoods....

> AMELIA: We went from Tommy to Steven. I'm not saying that Steven is bad and Tommy is great, but we went from one extreme to another. Tommy was very outgoing and very involved. His kids went to Garner [public schools] and he was always at things. Steven is the complete opposite. His children are home-schooled, and he doesn't go to school events where people from the church would be. He's not visible at all. Outside the church, he's not visible. Our kids need that, especially in the youth department....

[As Amelia speaks, the faint sound of the automatic garage door opening is fol-lowed by the sounds of a turning door knob, a squeaky opening of a door, a shutting sound, and footsteps. Will, Amelia's husband and deacon at the church, enters the kitchen, interrupting the interview.]

> AMELIA CONTINUES: Our new preacher Steven is a very godly man. He's very close with God. He is a fine addition to our community. [Amelia pauses and looks at the interview guide.] Do you want me to talk about what we do after church now?

We think voice in the agentic assemblage, not to emphasize the individual voice of the speaker in the research artifact, but rather to draw attention to the move-ment, or the agential force, of all sorts of voices (human and otherwise) that attach in an agentic assemblage to mark new territories and to create new becomings and different conceptions of voice. Such marking provides an in-between place for entering the territorial assemblage to see how voices in an assemblage are made, what they make, and what they do.

Amelia contrasted the new preacher, Steven, to the former preacher, Tommy, in her Baptist church when she said,

> I'm not saying that Steven is bad and Tommy is great, but we went from one extreme to another. Tommy was very out-going and very involved....Steven is the complete opposite.... He's not visible at all. Outside the church, he's not visible.

When voice is approached as *one* element in an agentic assemblage instead of as being spoken by a subject, a different conception of voice and purpose is made possible; Amelia shifts her description of Steven in the assemblage as other agents exert forces and pressures: "Our new preacher Steven is a very godly man.... He is a fine addition to our community." With this example, we assert that an individual subject, in this case, Amelia, does not merely "change her mind" about what she is saying but rather shows that such "dynamism is agency" (Barad, 2007, p. 141). This dynamism is not reduced to Amelia's agency but expanded to include the distributed agencies of the church, the community, Steven, the presence of a recording device, a garage door, foot-steps—an assemblage that produces an "ontological performance of the world in its ongoing articulation" (p. 149). According to Bennett (2010b), "there is not so much a doer (an agent) behind the deed ... as a doing and an effecting by a human-nonhuman assemblage" (p. 28, emphasis added).

What, then, is this effecting? What does an assemblage *do*, when it acts as a "heterogeneous 'team' of components"? (Bonta & Protevi, 2004, p. 56). To recall, an assemblage is a *process* of arranging or fitting together, and these arrangements create new ways of functioning:

> An assemblage emerges when a function emerges; ideally it is innovative and productive. The result of a productive assemblage is a new means of expression, a new territorial/spatial organisation, a new institution, a new behaviour, or a new realisation. The assemblage is destined to produce a new reality, by making numerous, often unexpected, connections.
>
> *(Livesey, p. 19, 2010)*

Specifically, mapping the function and effects of the assemblage is to see how a territorial field is both made (through repetition) and unmade (deterritorialization) through the rhizomatic connectivity of assemblages. Bonta and Protevi (2004) explain that "one should look to the openness to novelty of each assemblage, the way it invites new connections with other assemblages" (p. 55).

All agents were on one assemblage and then (re)distributed as other agents from another assemblage, an intra-action to take *everything* elsewhere, to fragment and re-organize to make a new territory. A particular arrangement of bodies, materiality, habits, institutions, and behaviors formed in that moment to enable and claim a territory of *critique*: Amelia, in the domain (assemblage) of domesticity is produced as expert from a working-class education and gendered upbringing, acts with the agents of/in the kitchen assemblage to make a new territory, one of critique. The assemblage produces her as mother who knows what is best for her children, who feeds those bodies, and for that instant, the territory of critique is formed, until the material force of footsteps reclaims the territory of patriarchy. The territory that was claimed in the kitchen assemblage is now fragmented and carried away. Will's footsteps organize to make a new territory, the territory of small-town conservative patriarchy where she is now no longer critical, but docile. Put another way, the function of this assemblage is to re-mark the territory of patriarchy—which also re-territorializes those things in the assemblage: "ownership" of a house, women's roles in the church and other domestic spaces—even what might be a habitual response to a hus-band's footsteps. In this other assemblage (of which Amelia's voice is only one element), a new becoming is articulated—articulated not by Amelia's voice but articulated *by the agential force of the assemblage*.

A reading of voice from a distributive perspective as described above seeks a reconsideration of how voice is produced that displaces simplistic notions of voice as only "spoken" by (human) research participants. We do not claim that the garage door "has" a voice, nor do a husband's footsteps "have" a voice—any more than we claim that Amelia "has" a voice. Certainly, that is one way to think about it. But from a posthuman perspective that necessitates a stance of distributive agency, those things are all part of an agentic assemblage that has an ontological force. Voice as emanated from a human subject is not primary here; instead, voice is just one *thing* entangled with other *things* that is becoming in a mutual production of agents, voices, becomings. In the posthuman, "material and human agencies are mutually and emergently productive [or constitutive] of one another" (Pickering, 1999, p. 373).

Voice in a Shifting Entanglement of Intra-Actions

Concomitant with Jane Bennett's agentic assemblage and distributed agency is a concept we take from Barad (2007), that of intra-activity. Intra-activity refers to the ways in which "discourse and matter are understood to be mutually consti-tuted in the production of knowing" (Lenz Taguchi, 2012, p. 268). Key to Barad's analysis is the agential realist understanding of "matter as a dynamic and shifting entanglement of relations" (2007, p. 224). For Barad, this understanding of agential realism "takes into account the fact that the forces at work in the materialization of bodies are not only social and the bodies produced are not all human" (p. 225). We take this then to move from voice as only produced by human bodies, to consider forces at work in the materialization of voice that is a shifting entanglement of relations—of bodies, mores, objects, unspoken thoughts, communities, spacetime, and so on. To open the concept of voice to "an outside whose determinations do not begin and end with the human subject" (Kirby, 2011, p. 15) but emerge in the intra-action (Barad, 2007) with other "things" in an agentic assemblage.

We recognize how from a posthumanist stance, agency is constituted as an enactment, not some-thing that an individual possesses, nor something that relies on a demarcation between human/non-human. In the research artifact with Amelia, we notated the multiple material and discursive forces (garage doors, gestures, speech, patriarchal discourses) not for the purpose of claiming difference, but for the purposes of, first, troubling those assumed demarcations and second, making transparent how/where/when voice is produced in the assemblage. Voice in the assemblage is continually made and unmade in the fits and starts and interruptions and connections that occur as Amelia is speaking, remembering, being jolted back into the present by the sound of the garage door—taken to a different assemblage, with a different function, and thus being reminded not to push back against the mores of her community. Those forces are drawn and connected from multiple milieus that defy linear, causal, and hierarchical conceptions of space/time/matter. We notice the agentic force of the entanglement of *things* as one assemblage is claimed (that of critique), then another entanglement of *things* organizes to mark a different territory with a different function. If we view voice as something materially produced in the agentic assemblage, then we understand Barad's (2007) assertion that intentionality is not the sole purview of humans but is "understood as attributable to a complex network of human and nonhuman agents … that exceed the traditional notion of the individual" (p. 23). This excess that is beyond the humanist subject is the agentic force or "thing-power" (Bennett, 2010b) of the material and discursive elements of the assemblage.

We see this at work in the research artifact. Both human and nonhuman agents are part of this acting network, this assemblage, this apparatus: Amelia, the small conservative town in which she resides, a working-class education that further inscribes traditional gender behaviors, small-town schooling that attempts to keep people in their place, the practices of a former preacher contrasted to the practices of the current preacher, Amelia's internal milieu of her own emotional terrain, sounds of a garage door and heavy footsteps, and so on. To further extend Barad's statement, agency can be thought as attributable to a complex network of agents "continually coming into being, fading away, moving around" (Pickering, 1993, p. 563), in a mutual and emergent production. It is this concept of agency as attributable to a complex network of agents that we work to think voice in a similar fashion (see Jackson, 2013 for a reading of data as/in/of the assemblage, or mangle). That is, we maintain that the voice that is claimed is *not attributable* to Amelia, but rather *bound to* the assemblage in which practices of etiquette; the presence or absence of spouses, pastors, children, old friends; status in the community; material surroundings;

institutional discourses; objects that are both symbolic and agentially real are all elements of the assemblage that are joining forces in this moment in an emergence of this posthuman voice. In the research artifact, there is no voice that stands alone to be extracted from the assemblage. Without the intersection and intra-action of forces in the agentic assemblage, there is no posthuman "voice," only words on a page. Therefore, voice in the agentic assemblage emerges as the "sum of the vital force of each materiality" (Bennett, 2010b, p. 24).

Conclusion: A Voice as *Doing*

> Expression is not in a language-using mind, or in a speaking subject vis-à-vis its objects. Nor is it rooted in the individual body. It is not even in a particular institution, because it is precisely the institutional system that is in flux. Expression is abroad in the world—here the potential is for what may become.
>
> *(Massumi, 2002, p. xxi)*

In this chapter, we have been thinking voice in the assemblage. Doing so has allowed us to posit voice in educational research not as something that *is*, but rather as something that *becomes* in an emergent intra-action with other agents in the agentic assemblage. That is, we moved away from positioning voice as simply spoken words emanating from a conscious subject in the case of Amelia, to voice as constituted in the entanglement of *things* (footsteps, squeaky doors, verbal critiques, institutional discourses, feminine bodies, male privilege). We illustrated with an example from a conventional research artifact in the form of a transcript how voice was bound to an entanglement of forces, where human and nonhuman bodies and accounts (matter and meaning) reconfigured voice as no longer attached to a humanist subject. In doing so, we have constructed a methodological move that enables a more provocative outcome than does an attempt to determine what Amelia's voice might mean.

We end, then, not with a further delineation of what voice is, or might become, but with a brief discussion of what might be made possible by thinking the methodological implications—of voice, of data, of analysis—in qualitative educational inquiry. Positioning voice, (and data, and analysis) in the agential assemblage presents three moves that we think warrant further attention in posthuman research practices in education.

First, positioning voice in the agentic assemblage takes us away from simple contextual analysis and forces us to pay attention to different spatial and temporal dimensions of voice. Doing so means that we recognize the agential cuts that are made, in the assemblage, and that force an attentiveness on the part of researchers to everything, *including* what is said. We can no longer think of Amelia's voice as separate and individual but only within the entanglement it immediately becomes and continues to become as it joins other enactments, other assemblages. The artifact or *voice* is no longer a story of Amelia, or of her experience but a mutual constitution produced by material and discursive forces that make the assemblage. Like Pickering's description of agents continually coming into being, and like Coole's description of agentic capacities that are contingently identified with individual agents, voice in the agentic assemblage is there right now in that moment, called up by the materiality of particular matter histories, discourses, signs, utterances, and so on. *It is not a matter of how a human voice articulates those things, but how the intra-action is an agential cut that assembles them and territorializes a space.*

Kirby (2011), in her reading of Barad, makes clear that in any entanglement, the entities do not exist separately or act independently of each other; "entanglement suggests that the very ontology of the entities emerge *through* relationality: the entities do not preexist their involvement" (p. 76). Based on this assertion, we treat voice as one such entity. Attention to these onto-epistemological entanglements make matter matter, enable an analysis that decenters the intentional human subject, and distributes agency among all of the *things* in an assemblage that is continually making and unmaking itself. Again, we assert that nothing is mediated; everything is made, including voice.

Second, if voice is no longer something to be retrieved to provide an account of a participant's experience, then data analysis is no longer a practice of providing a representational account. This move of intra-acting with the territorializations of the agentic assemblage is a practice that not only pushes against (over-)simplified treatments of voice-as-data but also the traditional, interpretive analytic strategies that tend to take for granted what voice "is" and what voice "does" in educational research. We think this makes possible an uncontainable voice in the assemblage that incites becomings and new ontological entanglements, new territories and de-territorializations.

Thinking voice, and analysis, thus provokes a different set of analytic questions when voice can only be thought in the agentic assemblage. Approaching analysis as a "plugging in" (Deleuze & Guattari, 1987; Jackson & Mazzei, 2012) that opens up potentialities, we advocate analysis as a process of developing analytic questions that seek the provisional emergence—of voice, subjects, agents—in the assemblage.[2] For example, returning to the conversation with Amelia and informed by the theoretical concepts that we have presented—agency, the assemblage, intra-action—we might ask:

- *How does Amelia intra-act with the materiality of her world in ways that produce different becomings?*
- *What are the points on the spectrum (of motherhood, small-town living, being a 'good' Christian, etc.) that co-exist and intra-act?*

Finally, thinking a posthuman voice in educational inquiry also requires a rethinking of interviewing. The research artifact presented above does not represent Amelia or her own, individual "lived experiences" but an assemblage of points on the spectrum (Amelia, other women like and unlike her, small towns, gender norms, disappointments, pleasures, place, and so on) that can produce a different set of questions and research practices that do not rely on a *single* source of knowledge. A posthuman voice in the agentic assemblage draws us to the present-ness and potentialities of that which did not unfold linearly via neat and tidy causality. That is, the elements in the assemblage (or entanglement) are not single sources of knowledge or agency—the footsteps did not cause Amelia to act any more than Amelia rushed to get her voice/critique "heard" before her husband arrived home. We need not separate the acts or agents to prove "some sort of mysterious connection among them" (Kirby, 2011, p. 77). Research artifacts are each *one* moment in *one* assemblage—or *one* particular entanglement. The point, for interviewing practices, is to notice how these elements—including voice—are joined in a particular way, "in the sheer wonder of the spacetime entanglement at work" (Kirby, 2011, p. 77).

Such a scenario certainly requires a de-centering and de-privileging of the method of interviewing in posthuman educational research. Thinking the practice of interviewing in the assemblage necessitates that as researchers we produce practices that are themselves entangled in order to allow the mutual emergence and production of forces to join other enactments and assemblages—kitchens, footsteps, husbands, hometowns. It means that we try to think "research as the machine that is a hub of connections and productions, with interviewing being just one of those connections" (Mazzei, 2013, p. 739).

In returning to the artifact from our conversation with Amelia, her words (her voice) do not exist apart from her or us, but in the assemblage, as a *voice without organs* (see Mazzei, 2013), as a knot of forces between the material and discursive. We follow Braidotti (2013), who writes, "The collapse of the nature-culture divide requires that we need to devise a new vocabulary, with new figurations to refer to the elements of our posthuman embodied and embedded subjectivity" (p. 82). It means that as educational researchers, we are challenged not only to devise a new vocabulary, but also to see what different possibilities might be produced for doing educational inquiry in the agentic assemblage.

Notes

1 This chapter was original published as Mazzei, L. A., & Jackson, A. Y. (2017). Voice in the agentic assemblage. *Educational Philosophy and Theory*, *49*(11), 1090–1098. It is reprinted here by kind permission of the authors and Taylor & Francis.
2 In our book, *Thinking with Theory in Qualitative Research* (2012) we engage Deleuze and Guattari's concept of "plugging in," as a *process* rather than a *concept*. One aspect of this process involves putting philosophical concepts to work via disrupting the theory/practice binary in order to see what analytic questions are made possible as they *think theory and data together*.

References

Barad, K. (2007). *Meeting the universe halfway*. Durham, NC: Duke University Press.

Bennett, J. (2010a). A vitalist stopover on the way to new materialism. In D. Coole & S. Frost (Eds.), *New materialisms* (pp. 47–69). Durham, NC: Duke University Press.

Bennett, J. (2010b). *Vibrant matter: A political economy of things*. Durham, NC: Duke University Press.

Bonta, M., & Protevi, J. (2004). *Deleuze and geophilosophy*. Edinburgh: Edinburgh University Press.

Braidotti, R. (2013). *The posthuman*. Malden, MA: Polity.

Coole, D. (2005). Rethinking agency: A phenomenological approach to embodiment and agentic capacities. *Political Studies, 53*, 124–142.

Deleuze, G., & Guattari, F. (1987). *A thousand plateaus: Capitalism and schizophrenia*. Minneapolis: University of Minnesota Press.

Hekman, S. (2010). *The material of knowledge: Feminist disclosures*. Bloomington: Indiana University Press.

Jackson, A. Y. (2003). Rhizovocality. *International Journal of Qualitative Studies in Education, 16*, 693–710.

Jackson, A. Y. (2013). Posthumanist data analysis of mangling practices. *International Journal of Qualitative Studies in Education, 26*, 741–748.

Jackson, A. Y. & Mazzei, L. A. (Eds.). (2009). *Voice in qualitative inquiry: Challenging conventional, interpretive, and critical conceptions in qualitative research*. London: Routledge.

Jackson, A. Y. & Mazzei, L. A. (2012). *Thinking with theory in qualitative research: Viewing data across multiple perspectives*. London: Routledge.

Kirby, V. (2011). *Quantum anthropologies*. Durham, NC: Duke University Press.

Lenz Taguchi, H. (2012). A diffractive and Deleuzian approach to analysing interview data. *Feminist Theory, 13*, 268–281.

Livesey, G. (2010). Assemblage. In A. Parr (Ed.), *The Deleuze dictionary* (revised edn.) (pp. 18–19). Edinburgh: Edinburgh University Press.

MacLure, M. (2009). Broken voices, dirty words: On the productive insufficiency of voice. In A. Y. Jackson & L. A. Mazzei (Eds.), *Voice in qualitative inquiry: Challenging conventional, interpretive, and critical conceptions in qualitative research* (pp. 97–113). London: Routledge.

MacLure, M. (2013). Researching without representation? Language and materiality in post-qualitative methodology. *International Journal of Qualitative Studies in Education, 26*, 658–667.

Massumi, B. (2002). Introduction: Like a thought. In B. Massumi (Ed.), *A shock to thought: Expression after Deleuze and Guattari* (pp. xiii–xxxix). New York: Routledge.

Mazzei, L. A. (2013). A voice without organs: Interviewing in posthumanist research. *International Journal of Qualitative Studies in Education, 26*, 732–740.

Pickering, A. (1993). The mangle of practice: Agency and emergence in the sociology of science. *American Journal of Sociology, 99*, 559–589.

Pickering, A. (1999). The mangle of practice: Agency and emergence in the sociology of science. In M. Biagioli (Ed.), *The science studies reader* (pp. 372–393). New York, NY: Routledge.

Ringrose, J. (2011). Beyond discourse? Using Deleuze and Guattari's schizoanalysis to explore affective assemblages, heterosexually striated space, and lines of flight online and at school. *Educational Philosophy and Theory, 43*, 598–618.

Snaza, N. & Weaver, J. (Eds.). (2014). *Posthumanism and educational research*. New York: Routledge.

St. Pierre, E. A. (2004). Deleuzian concepts for education: The subject undone. *Educational Philosophy and Theory, 36*, 283–296.

6

WONDERING IN THE DARK

The Generative Power of Unknowing in the Arts and in Qualitative Inquiry

Liora Bresler

The invitation to write this chapter came at the right moment, just as I was reading students' final papers for my last qualitative class whose closing session was the day before. With the concluding chord of a long and rewarding institutional career, I was moving (for once, deliberately choosing to move), into a zone of unknowing. Without tight structures and schedules, tasks and commitments, I did not know who I would be. Norman Denzin and Michael Giardina's request to write a manuscript based on my International Congress of Qualitative Research (ICQR) talk on the "Generative power of unknowing," I knew(!), would support the investigation of my own actual experiences of unknowing with their multi-layered qualities, emerging meanings, and rewards. Indeed, this chapter, coinciding with and echoing the emerging chapter of my life, facilitates moving and being moved, back and forth, between the experiential and the verbal, both essential poles of my academic writing. Attending to the small bread-crumbs en route to the unknown, it traces the way to the home I am leaving as well as charting a fresh path.

The concept of unknowing can be too quickly classified as disgraceful ignorance or laudable beginner's mind. Yet, the *experience* of unknowing has a richness, vibrancy and range of conflicting emotions that defy neat categorization. Having been intrigued by the generative power of unknowing in qualitative research for some 35 years, grappling with its forbidding aspects, as well as with its affordances and possibilities, I am now settling into an intentional exploration of it, rather than rushing to the knowing layer. The arts have been major forces in shaping and establishing my identity as knower, then unknower. This chapter centers on the interplay in the arts and in qualitative research, between knowing and unknowing—forces that at times resist each other as incompatible opposites, but more often support each other towards greater depth and fulfillment, with vital implications for the conduct of qualitative research.

Some unknowing may feel heavy, dark and threatening. The unknowing I embark on now feels like a musical convergence of sounds with their distinct dynamics, timbres and evolving patterns beckoning me to work with what develops in ways that are newly responsive. Composing this chapter of life involves tuning in to an inner self with curiosity rather than leaning heavily on existing knowledge and habits. In this process, my personal crossroad seems to coincide with a well paved methodological crossroad, to branch in unexpected direction of connecting and being.

Knowing: Music

Music was my home base, before conscious memories. By age three, when my parents bought a piano, I sounded notes, traced melodies of Israeli folksongs, harmonies and rhythms, and soon sharing those with the audiences of my preschoolers and friends of family who gathered to sing together. I knew the songs through intimate connections of ears and fingers, basking in the glow of a small community united by music.

Starting formal piano lessons at nearly six on the condition that I would stop improvising (which, my teacher explained to my parents, contaminated my technique), I acquired acceptable fingering, knowledge of notation, and some music theory. Knowing a classical piece involved eyes, ears, mind and body. It started with transferring musical notation into sound, honing technical skills, and imprinting those skills on the memory of the body. (It took the writing of this paper to note the apt choice of words in the idiom "knowing a piece by heart.") Musical knowledge was framed as mental and auditory rather than embodied, with fingers serving as tools. This common view did not fool me: I realized the essential role of the fingers since it was my fingers that knew what to play when my mind went blank, as it sometimes did in the dreaded annual recital, so different from the carefree playing of folksongs.

I perceived the learning of music, divided neatly into separate disciplines (theoretical versus performance), as overcoming the infinite terrain of not knowing by means of acquiring new repertoire, gaining analytic tools, mastering technique.[1] Music theory, embedded in sound, presented its own kind of knowledge and beauty. I was captivated by the architectural structure of the circle of the fifths with its inner logic and intricate relationships. Analysis of musical pieces was often an opportunity to discover little universes of intelligent design. Teaching piano to both children and adults as a young child and teenager, and later, music theory at the Open University in my early twenties, provided opportunities to share those gems.

The layered nature of knowing in music—verbal and auditory, concrete and abstract, explicit and implicit—was foundational to the layered nature of unknowing/knowing. Music theory—including scales and harmonic progressions, and musical forms with their thematic structures, period-appropriate styles, rhythm and dynamics—is at once auditory, symbolic and conceptual. The level of integration between the theoretical and experiential aspects of music ranges from fusion (for example, in the analytic harmonic progressions and counterpoint or period stylistic practices) to loose connection (music history's focus on the lives of composers). These understandings, while tacit in my early years, have moved to the explicit, verbalized level with my personal and scholarly explorations of knowing.

Pedagogy proved to be as essential to cultivating relationships as is the subject matter itself. If music theory was playful, piano performance, highlighting technical proficiencies, was strict, seeming to allow no hint of personal expression and no meandering, which took the life and pleasure out of my playing. The aim of getting closer to the "composer's intent," a key aspiration in the culture of classical music performance, felt more like reproducing than creating. With certificates from the Conservatoire, a BA in Piano Performance and Pedagogy, and a Master's in Musicology, I developed the reassuring identity of an expert with its feel of solidity. This identity came at a price: the abandonment of improvisation and dabbling in composition, the loss of musical exploration and ownership. My earlier delight in playing music was replaced with a sinking heaviness during lessons and minimal home practice. The energy of the audience in concerts (unlike that of my forbidding teacher) still created occasions for the spark of connection. Nevertheless, my acute awareness of falling short of my own vision, and my resistance to practice, brought my piano-playing to a slow ritardando. Rather than rebel, I gave up ownership and largely withdrew from performing classical work. I still continued to play communal folksongs when the opportunity arose, with the same joy, and skills, of my three-year old self. Importantly, listening to classical music maintained its power, connecting me to what I experienced as the sublime, as well as to myself.

The tension between a discipline experienced versus a discipline studied was not unique to music, of course. When I did a university double-major in philosophy, I found most assigned readings to be dry, anesthetic, and disconnected from lived experience. In contrast, the practice of logic, similar to music theory but considerably more adventurous, felt playful. I discovered, on my own, existential philosophy and humanistic psychology, not part of course offerings. Sartre, Camus, Frankl, Maslow, and Rogers spoke to me directly and forcefully, connecting with my inner self, experiences and queries.

While musicology courses were as dry as those in philosophy, their musical denotations made them more palatable than abstract language. I was engrossed in my master's thesis (longer and more time-consuming than my later PhD), examining how the Israeli music I grew up with was shaped by the history, ideology and missions of the 1930s and 1940s. The process of inquiry had vitality and resonance. Its surprises (for example, that the same principles and ideologies I found in both "serious" and popular/folk music were also prevalent in the artistic domains of visual arts, dance, and theater, pointing to larger patterns that I had not realized existed), expanded my knowing.

Unknowing: Educational Research

It was against this background that I was plunged into unknowing. A move to the United States meant being pulled away from my personal, social and professional networks, including a treasured position directing musical concerts at the Tel-Aviv Museum. Unknowing felt disorienting rather than generative. Off my familiar track, in a totally new habitat (Bresler, 2016), my sole aim was staying afloat. In this transition, decontextualized bits and pieces from my philosophy and music background helped launch me into the discipline of education. Ernest Gombrich's *Art and Illusion* (1956), a book I read (and loved) as an undergraduate in philosophy was the first reading in a Stanford seminar led by Elliot Eisner, which I happened to visit. Enticed by Elliot's enthusiasm, and his contagious ability to make aesthetics come alive, I kept coming back throughout the semester. It was a surprise, a week after the seminar was over, to be offered a research assistantship with Elliot.

When I confessed to not knowing anything about education, Elliot was undeterred, reassuring me I would be just fine.[2] My complete ignorance of the fields of curriculum, education and qualitative research meant that I needed to draw on other sources. During that first fieldwork in an elementary classroom, I found musical dimensions—temporal form, rhythm, orchestration, melody, counterpoint and dynamics—to illuminate classroom life (and the wider array of lived and composed experiences, including the experience of giving talks, creating syllabi, and editing books).

I should not minimize the role of serendipity in this constellation. Eisner, my recruiter and gate-keeper, was, unbeknownst to me then, a passionate advocate of art criticism as a model for educational inquiry (Eisner, 1979), and my lenses fitted well within this model. It also helped that my area of knowledge, music, extended Eisner's visual arts-based model, in attending to another "form of representation" (Eisner, 1982, p. 1).

Teachers are central in ushering newcomers into a discipline. Eisner's generous welcome, aesthetic thrust and engaging course materials; Decker Walker's inquiring mind, intellectual integrity, and impressive deliberative style radiating in a curriculum class I took with him that first year of doctoral studies in 1983; and Nel Noddings' focus on existential theology, complete with her caring pedagogical presence, made this transition rich and attractive. While I appreciated the vitality and relevance of course readings, it was the prolonged engagement with fieldwork and interpretation that was transformative, establishing a new relationship to knowing/unknowing. The grappling with what I observed in classrooms and heard during interviews were spaces of unknowing, where ideas took time and care to take shape. The processes of observations and semi-structured interviews drew on practices and sensitivities I had cultivated throughout my life, now formalized and systematized. Likewise, data analysis invoked identification of motifs and patterns—a process that, at a deeper level, remained intimately familiar from years of musical analysis. The unknowing involved in observations, interviews and analysis was different from the not-knowing of vocabulary, bodies of knowledge and skills sets.

The stance of a non-participant fieldworker facilitated the cultivation of "aesthetic detachment" (Beardsley, 1983) from what I observed. That detachment was connected and invested, not unlike the one I had experienced in listening to music and looking at art. The juxtaposition of distances involved in research—close observations and interviews, along with the removed perspective of data analysis—supported "seeing more" in what I studied, moving from the implicit and experiential to the explicit and verbal.

A fourth course I took that first year in cultural anthropology, with George and Louise Spindler, introduced the notion of the researcher as an outsider who aspires to understand insiders' knowledge, just as I did in my fieldwork. The narratives of "strange" and "familiar" settings (Spindler, 1982) and the compelling conceptualizations generated by this back-and-forth of "strange/familiar" echoed the interplay of my home/new culture, intimate/distant, and helped me grapple with classroom observations in my research assistantship, as well as with my own evolving life.

The context of educational research in those early 1980s provided a relevant framework for unknowing. Elliot Eisner and Decker Walker supported my qualitative inquiry experience from the very start of my doctoral education, accepting emerging issues as part of their (distinct) personalities and interests. They belonged to a broader group pioneering curriculum research. In the aftershocks of the turbulent 1960s with their civil rights protests, women's movements, and student riots, there existed little prior educational research (or practice) attending to these marginalized voices, which signaled a profound deficiency. A recognition that the knowledge produced by educational research was often irrelevant to pressing educational issues alerted curriculum theorists to the inadequacy of their researcher goals and corresponding tools and prompted them to turn to anthropology instead of positivist psychology for new paradigms and methodologies. Joseph Schwab, a major figure in curriculum theory, claimed that the field of curriculum was "moribund" (Schwab, 1969) because it was disconnected from practice. New and innovative notions of curriculum included the *hidden* curriculum (Dreeben, 1967; Jackson, 1968) and the *implicit* and *null* curricula (Eisner, 1979), ushering important curricular aspects that had not been considered in established theories and practices. Recognizing that there were several—sometimes connected, sometimes parallel—curricula, Decker Walker discussed curriculum as subjects, activities, intentions, and experiences (1989), advocating deliberations among different stakeholders, while

John Goodlad highlighted ideal, formal, operational, perceived and experienced curricula (Goodlad, Klein & Tye, 1979). An awareness of multiply interpreted, layered concepts supported the exploratory nature of qualitative research.

I did not know any of this when I was a doctoral student. Some understanding came with my expanding familiarity with the broader context of education.[3] But I sensed a lack of rigidity (as compared to the 1970s world of musicology[4] that felt utterly well defined and stifling) in the openness, freshness and opportunities of educational research. The lack of reliance on a narrow theory invited a responsiveness in fieldwork and freedom to follow enticing issues. Working with field experiences and carving a research path encouraged cycles of interactive questions.

The transition to a new country and discipline, both of which I found welcoming and hospitable, involved busyness in acquiring new competencies and knowledge, cultural and disciplinary. The courses in my doctoral program provided a direction, map and sequence. *But the quest for knowing more was founded on realizing the generative power of unknowing as an underlying formative experience.* The epiphany (Denzin, 1989) of unknowing spawned some new themes and methodologies (Bresler, 2006). Working with the unknown, expecting it, lingering with it, aiming to evoke curiosity even when threatened by it, all became important aspects of my research process. The interplay between knowing and unknowing assumed some ease, even agility.

Glimpses into Buddhist Worldview and Research

Just as it was the experience of fieldwork that lured me into educational research, it was the experience of a meditation retreat that lured me into Buddhist worldviews. In June 1987, a week after I deposited my doctoral dissertation with (adorable) twin toddlers at home, and ready for a pause, I enrolled in a silent meditation retreat. Though it took me years to deepen my understanding of the Buddhist philosophy underlying the practice, the fundamental quality of unknowing was tangibly present in those five days of no-talk, no-eye-contact experience. The practice was not "fun" or relaxing, as people often assumed when I mentioned the retreats (which would become a regular part of my life), but it felt meaningful.

If cultural anthropology and qualitative methodology provided a professional framework for research, my growing interest in Buddhist worldviews became a rich source of insight into the underlying process of perception and interpretation. Buddhist practices highlight inward observation and awareness of the fluidity of phenomena. Juxtaposing focused concentration with spaciousness increases clarity (Goleman & Davidson, 2017). The notion of "beginner's mind" (Suzuki, 1970) fit well with the freshness of unknowing. Among the several Buddhist-related stories I told my students, one describes Nan-In, a Japanese Zen Master who continues pouring tea into the full cup of his Professor of

Philosophy guest. When the Professor becomes alarmed by the overflow, the master suggests that until the guest empties his head, there is no space for new knowledge. The fullness of expertise with all its attachments, I remind myself, can come with a price of not knowing that we don't know.

Emptying one's head, letting go of knowledge, is no small task, especially within a scholarly culture that centers on knowledge. Moreover, a life of consuming and creating knowledge testifies to my deep academic commitment. Rather than get rid of knowledge, I aspire to be mindful of productive ways to work with it—noting its wise and unwise uses, boundaries, and limitations—and to open a space for unknowing as enabling and expanding.

Developing Academic Practices

With a PhD in hand, I came to work with Robert Stake, whose writing I admired in graduate school. A pioneer of qualitative research in education, he coined the concept of *responsive evaluation* (1967), advocating that evaluators respond to a setting's participants and issues. Bob, who proved to be as inspiring in mind and heart as in his writing, invited me to join his project of arts education curricula in the United States, funded by the National Endowment of the Arts.

Having been raised in a different culture with its distinct value systems and practices, I was conscious of my outsider frame of reference and the importance of acknowledging my unknowing for productive fieldwork. The questions I initially posed in proposals were good enough, grounded in contemporary scholarship and my already articulated interests; however, the ones I garnered in the field were often more authentic and vibrant. Accordingly, prior to fieldwork, I jotted down some queries under the watchful eyes of Bob, who wanted to make sure I knew how to formulate research questions, but as I acknowledged to him, these questions were just a starting point: the real questions would surface in the field through interaction, evoking existing even if unformulated wonderings.

Our initial foreshadowing questions had to do with the schools studied, their customs, curriculum and aesthetic context (Stake, Bresler, & Mabry, 1991, 5). The *emerging* questions and themes revealed puzzling little mysteries. I could not, for example, comprehend why the classroom teachers whom I observed practiced teaching one way (teacher-centered pedagogy; relatively uniform-looking arts products) and espoused another (student-centered pedagogy; art as self-expression and individuality). It took me months to realize that context—the culture and broader practices of schooling—was paramount to the teachers' practice (Bresler, 1992). It took me several more months to discern similar types of discrepancies present in my own life. A particularly humbling one was my obvious joy when my eight-year-old children engaged in making a (stereotypical) Hanukah craftwork at a party, exactly the kind I was condemning in a newly published journal article.

Another methodological issue highlighting unknowing as generative force was the notion of "interpretive zones" (Bresler et al., 1996; Wasser & Bresler, 1996), highlighting an attuned presence in teams' joint interpretations, akin to playing chamber music. Similar in mindset to interviews and observations, the focus was on relating to others' knowing (addressing our blind and blank spots), highlighting and complementing our individual and collective unknowing.

My grand research failure happened (luckily) when I was already a veteran full professor, after years of diverse projects investigating arts education in ordinary and extraordinary schools. I had embarked on a study exploring aesthetic experiences in a cultural performing center, with the passion and commitment one has to a very special setting, the refuge and haven that my campus' Krannert Performing Arts Center has been for me since I came to Illinois. I headed a team of competent and interested research assistants. The failure was as monumental, I felt, as the historic sinking of the Swedish ship Vasa in August 1628, just as it left the Stockholm Harbor. Not unlike the Vasa's mighty, cumbersome canons, my knowing was too weighty and just as futile. Like King Gustav Adolf, who was attached to the ship's splendor, I, too, was attached to my reverence for the Center, which functioned as my "artistic church." There was plenty of lingering with the institution and huge amounts of data, meticulously organized: observations of a wide range of performances; interviews with artists, audience and Krannert staff; program notes and other archival materials. The study sank after an extended data analysis,[5] without publications.[6] My extensive literature search and theoretical understanding of all this did not help. I was not able to let go of what I "knew" I'd discover to find out what I didn't know. This experience confronted me with the price of attachment to knowledge.

The Pedagogy of Cultivating Unknowing: Lingering with Artwork

The same qualities that were vital to me as a researcher propelled me to grapple with the nature of a qualitative mindset around which to build my research methodology courses. Aiming to sensitize my students to the generative power of unknowing, I have grappled with the "teachability" of cultivating unknowing in the service of intensified perception; a taste for asking genuine and generative questions; and the ability to respond to what we encounter with presence, intuitive sensitivity, and agility. This, I found, is where the arts can serve as catalysts for the learning process. (The arts world, as I have experienced first-hand, provides seductive traps for expertise. Accordingly, my use of aesthetic-based pedagogies involves engagement with artwork for audiences who are not experts in those artistic media, e.g., the use of music, visual art, and dance for researchers in general education; the use of visual arts for musicians.)

Recognizing the complexity and ambiguity of the arts, I drew on aesthetic materials for my research courses to invoke a polyphony of meanings, letting go of ready-made knowledge to allow space for fresh discoveries. For example, an initial assignment requires an extended time for viewing two artworks, 30–50 minutes with each. This assignment aims to generate an unfolding of *perception* as a richer ground for *conceptualization* of what we observe, creating a bi-directional relationship between them. It acknowledges the dual roles of inner and outer, facilitating awareness of subjectivities (Peshkin, 1988), including values, lenses, and evolving emotional and intellectual responses (for the detailed assignment, see Bresler, 2018).

It is only after lingering with the artwork, a pause within the typical reading/ writing academic tasks, that students are asked to examine contextual sources. Contextual, archival sources provide a layer of established scholarly information and expert interpretation. Knowledge and expertise, foundational to academic culture and its *raison d'être*, orients us to the traditions and wisdom of disciplines. They provide structures and tools to support intellectual explorations and create new knowledge, adding to, but not replacing students' response. The makeup of students from diverse interests and disciplines supports pausing from the conventional activities of reading and writing to listen to inner voices as well as to voices of other students, cultivating unknowing to engage conceptually, affectively, and authentically with unfamiliar materials to create fresh connections and catch a glimpse of our respective unknowing (for examples of students' work, see Bresler, 2018).

A Momentous Shift: Unknowing in Creating Movement

Even as I was practicing (and professing) unknowing in the conduct of research, I held on to my expectations of expert knowledge in the arts, a key aspect of my musical enculturation. My first deliberate encounter with artistic unknowing came when I asked the then new department head of Dance at Illinois, Jan Erkert, to address dance in my graduate course on aesthetics and curriculum. Jan engaged my students and me in Authentic Movement (Pallaro, 1999). When class was over and we "debriefed" in her office, Jan, who was a newcomer to campus, asked me if I would be her partner in Authentic Movement. I found her session powerful, and the invitation enticing. Obviously, I was totally lacking in knowledge or skills of movement, but the power of expression I had just experienced in this class was irresistible. I was thrilled to embark.

Jan and I have been doing Authentic Movement on and off for these past 12 summers. The practice proved to be a source of insight (and delight!) in ways I could not have anticipated. Here, it was not the mind commanding the body, bidding it to drill and do its tasks, as it was in my formal training in classical music. Rather, the mind had to learn how to be quiet as the body led. I was not responsible for creating anything beautiful or even interesting, nor aiming to sound intelligent or coherent. Jan, in the role of a witness, possessed an extraordinary ability to observe and articulate, often helping me put words to my

kinesthetic knowledge. What emerged in these sessions were qualities of experience that were body-based rather than explicitly conceptual. I realized first-
hand that art-making can indeed be a way of "ordering the self" (McConeghey,
2003, p. vi).

Spontaneous rather than planned, the process of exploring resonant movements, combined with the gift of a perceptive, eloquent witness, allowed me to
interact with emerging experiences and process them, as I was changing shape.
My current transition to Emerita as I am writing this chapter evidences new,
lighter movements; conserving energy and exploring inner forces; working with
the density of space as compared to the solidity of walls, bar and floor; and more
generally, venturing beyond my past repertoire of movements.

Different disciplines address different angles, as my very recent discovery of
Sondra Perl's work on embodied knowing within the discipline of creative
writing attested yet again. Perl's questions reverberate: "What happens when we
reach the edge of our thinking? From where do new thoughts come? How do
we go from what we know to what we don't know? Or from what we don't
know to what we know?" (2004). My experience of Authentic Movement res-
onates with those questions, placing the body as a central source of insight and
processing.

Unknowing in *Making* Art

While dance and creative movement belonged to my "null curriculum," visual
arts education had been a requirement in my elementary school. The focus on
drawing skills rather than expression left me cold. Only in my late teens did
attending art films, theater, and dance performances become important. I discovered art museums in Europe (even though there were plenty of art museums
in Israel), appreciating the contemplative space they afforded and the richness of
images and energies in the artworks. I responded early to the artworks of van
Gogh, Munch, and Miro (for example), recognizing that they spoke to and for
me, portraying a range of qualities, from angst and turmoil to the spirited and
whimsical. Reading about artistic movements gave me a framework for understanding artists' exploration of colors, lines, and themes and their play with
imagination, going beyond established boundaries and conventions. Sparked by
museum visits, I felt invited to mentally envision and perceive the world in
ways that were fresh, expressive and playful. In my role as audience (sometimes
aesthete) in the arts, I have brought some of these qualities to my home
environment, playing with colors, and shapes; small and big ways of organizing
my house and my life.[7]

It was a conversation with Peter London, a colleague whose talks in the
National Art Education Association and writings (London, 2003, 2007) I found
inspiring, that kindled an interest in his five-day course "Drawing Closer to
Nature." Apprehensive about my total lack of technical skills but reassured by

Peter's conviction, as strong as my own opposing one, that one needed no skills to attend this course, I plunged into it during the summer of 2017. Looking closely at nature (a process quite different from looking closely at artworks!) was radically new and unexpectedly rewarding. Peter built a sophisticated, pedagogically rich syllabus, with diverse assignments that aimed to connect the outer world with our inner selves. The course provided a space to create and communicate, sometimes with the whole group, sometimes with a partner, to observe, reflect and share observations.

Peter's powerful and inspiring condensation of Abraham Joshua Heschel's essay "Man Is Not Alone" (1951), is grounded in unknowing, as evidenced by this powerful and inspiring statement: "Everything and everyone you have ever come to know is actually different and more and better than you have come to know it as, will ever know it as, and can ever know it as, including your own Self." This speaks to the mystery of encounter, not closing off past, present, or future.

The following summer I attended Peter's course again, exploring new materials with their affordances, and deepening qualitative mindsets. Finding myself excited about some of my creations in ways that I did not anticipate, I let my hand and eye guide me, with my mind in the back seat, not intervening. As Peter observed, my artwork had incorporated musical elements of melodic lines and staccato. (My husband, Yoram, responding in an email to the artwork I shared with him, wrote independently: "It is like music—with evolution, echoes, undertones, reprises, chords, dark notes and brightness—but only two dimensional and colorful rather than temporal and auditory. As the eye traverses the different parts of your piece, it feels like listening to music.")

I learned that the artwork often anticipated unformulated experiences of past and present: feelings about the past, values of what nurtures and expands. I experienced first-hand an earlier understanding facilitated by juxtaposing distances: here, the close, intimate one of making of art; the "farther" one through observing what I made; and, importantly, listening to others' observations.[8]

Unknowing in Academic/Personal Writing

Alan Peshkin's (1988) invitations to be aware of one's subjectivities as they unfold in research, and Barbara Myerhoff's masterful portraiture of her researcher self in process (Myerhoff, 1978), have motivated me to note and reflect on this personal/professional interplay. My references to this occurred in teaching and conference presentations as part of trustworthiness in research. But it took me reading the work of narrative, folklore, and children's literature researcher, Betsy Hearne, to stimulate my next shift. In A *Narrative Compass* (Hearne, 2009), Betsy referred to stories that inspire and shape professional identity. Her notion of the "story-tuned" scholarly self inspired me to venture into a more explicitly professional/personal writing style. Betsy's work (Hearne,

2015, 2017) illustrates first-hand the power of integrating the personal with the professional in ways that are powerfully aesthetic, vigorous and inviting. It has helped having her in the role of a "witness" to my emerging writing, listening attentively, nodding with approval, even enthusiasm, when I was doubtful, providing me with just the right word, all with profound wisdom and generosity. Betsy's writing and witnessing served to strengthen my inner narrative and artistic compasses.

Writing requires a community. A community of intellectual kindred spirits is a precious gift. I have been grateful to have Norman Denzin create and recreate, year after year, the very special, immensely stimulating intellectual space of the International Congress of Qualitative Research, where I first presented these ideas. Autoethnographic writing shares with Authentic Movement and with making art a transformation of knowledge that is personal and visceral, as well as verbal and conceptual, exploring inner landscapes in the service of identifying larger issues. The important work of Carolyn Ellis on autoethnography (e.g., Ellis, 2009), offering methodological possibilities and compelling examples, as well as her joint work with Art Bochner (Bochner & Ellis, 2016), have extended my notions of the personal and professional. Art Bochner's autoethnographic work, including his book *Coming to Narrative* (2014), highlights processes of becoming, encouraging me to go back to my emerging self as a scholar, the readings and experiences that propelled and sparked me.

The Crucial Need for a Pause

Crucial in the journey from knowing to unknowing is an extended time to immerse oneself in perception and process. This extended time, a prolonged engagement (Lincoln & Guba, 1985), is indeed the hallmark of qualitative research. Its equivalent in creative acts is incubation; and in art appreciation a lingering caress (Armstrong, 2000). These terms imply a pause, time "off" from the mundane and the usual pace.

Perl (2004) suggests that "new ideas, or fresh ways of speaking, thinking, and writing will come to us if we pause and wait patiently, if we allow it to open." Building on Eugene Gendlin's notion of *felt sense* (1981), Perl discusses how this idea helps us clarify the dynamic relationship between thought and language. By directing attention to our felt sense, she suggests that "we establish a living connection between what we sense and what we know or between what we sense implicitly and what we state explicitly" (Perl, 2004).

A pause is essential for staying with unknowing in the transition to explicit knowing. Sometimes it takes several years after fieldwork for issues to emerge to a "knowing" stage. My dissertation, for example, examined the integration of computers into a music theory class, its possibilities and impact on the curriculum. Not until my dissertation was deposited did I become aware of another issue generated in that fieldwork that has been calling me with increasing clarity and urgency. Deeper than the specific technology used, this questioning underlay the structures of schooling: what does it take to be successful students in a class? In the music theory class featured in my dissertation, students who were engaged in musical activities outside of class (sometimes quite accomplished) received lower grades or dropped out, while students who were sometimes literally tone-deaf received A+ (Bresler, 1993). This startling observation can be understood by examining what tests measure and what knowledge is valued in music theory.

Coda: Tuned Listening

American dancer and choreographer Agnes De Mille is quoted by Buddhist teacher Pema Chödrön in a broadly applicable observation:

> Living is a form of not being sure, not knowing what next or how. The moment you know how, you begin to die a little. The artist never entirely knows. We guess. We may be wrong, but we take leap after leap in the dark.
>
> *(in Chödrön, 2013, 1)*

Methods and theories can serve as an external compass, by their nature belonging to an area of knowing. Carving our own paths of discovery requires an *internal compass*. The inner voice, often embodied and sensed rather than recognized and verbalized, requires tuned listening. My own research processes and pedagogies aim to support that tuned listening, opening a dialogue between the encountered and the self. Being a qualitative researcher shares with arts, and with life, that exploratory leap.

This processing aligns with John Dewey's long-ago prompt for "perception" instead of the automatic-pilot mode of "recognition" (Dewey, 1934, 52). Recognition, and habituation (Goleman & Davidson, 2017), obviously have their important uses. This chapter addresses what I see as the missing mind-state of unknowing: Underlying perception in all its openness is an essential part of the journey towards expanded visions and understandings.

Indeed, I have come to see the unknown as a constant companion through life. It is there in every new course I take or teach. It is present in travels, and, when I am tuned to it, in my supposedly familiar surroundings. It is present in meeting new people, as well as in encountering those I presumably know well, including myself. With the intent of communicating to you, the reader, on your own journey and with your own possibilities of resonance and variations, this chapter has indeed served me as a fulfilling space of exploration, witnessing the intimacy of my own "in-here" metamorphosing into "out-there."

Acknowledgment

Heart-felt thanks to Art Bochner, Carolyn Ellis, Alma Gottlieb, Eve Harwood, Betsy Hearne, Peter London, Julia Makela, Kitty Schmidt-Jones, and Decker Walker for reading early drafts of this chapter and for their insightful comments.

Notes

1 Much not-knowing, of course, stayed as such: I have never learned, for instance, to play dodecaphonic or aleatoric pieces and never mastered Schenkerian analysis.
2 Only writing this do I understand that through his model of expertise ("connoisseurship and educational criticism," Eisner, 1979) as a metaphor for educational inquiry, Eisner must have believed in the power of "unknowing."
3 The complex, sometimes intangible, and elusive nature of knowledge was identified earlier in various disciplines. For example, philosopher Michael Polanyi observed in his key, related concepts of "personal knowledge" (1958) and "tacit knowledge" (1966), that we know more than we can tell. This kind of knowing, tacit foreknowledge of yet undiscovered things, is embedded in the body rather than in language (Perl, 2004). A more historical search for conceptions relating to unknowing could include Freud and Jung, among others, but this would require a whole different manuscript about important differences among intellectual traditions, worldviews, and functions of these related concepts.
4 Musicology, too, was affected by this movement, several years after I left it for education.
5 Using, first (and last) time, sophisticated computer software.
6 Except one, based on youth performances attended by school groups. I noted, with chagrin, teachers' lack of identification with the musical experience, and their predominant disciplinary role (Bresler, 2010).
7 A recent example is my daughter's observations of family dynamics during her visit, noting my insistence that "the butter likes to be next to the cheese" in the fridge. "Why do you anthropomorphize, Mom?" she asked. "Oh, they both spent time in the cow's belly!" Ma'ayan describes the scene later in an email: "I keep pressing, and mom eventually retreats to her usual vocabulary: music. 'It's like two instruments in dialog. They're not really in dialog, but you lose something in listening if you don't yield to that perception.'" (Ma'ayan Bresler, September 5, 2018).
8 This was an insight I was espousing earlier in my courses and in my writing, (Bresler, 2014) but here it hit me with intensified freshness.

References

Armstrong, J. (2000). *Move closer: An intimate philosophy of art*. New York: Farrar, Straus, & Giroux.
Beardsley, M. (1983). Aesthetic definition of art. In H. Curtler (Ed.), *What is art education?* (pp. 15–29). New York: Haven.
Bochner, A. (2014). *Coming to narrative*. London: Routledge.
Bochner, A. & Ellis, C. (2016). *Evocative autoethnography: Writing lives and telling stories*. London: Routledge.
Bresler, L. (1992). Visual art in primary grades: A portrait and analysis. *Early Childhood Research Quarterly, 7*, 397–414.
Bresler, L. (1993). The social organization of achievement: A case-study of a music theory class. *Curriculum Journal, 4*(1), 37–58.

Bresler, L. (2006). Embodied narrative inquiry: A methodology of connection. *Research Studies in Music Education, 27*, 21–43.

Bresler, L. (2010). Teachers as audiences: Exploring educational and musical values in youth performances. *Journal of New Music Research, 39* (2), 135–145.

Bresler, L. (2014). Research education in qualitative methodology: Concerts as tools for experiential, conceptual and improvisatory pedagogies. In C. Conway (Ed.), *Oxford handbook of qualitative research in American music education* (pp. 608–636). New York: Oxford University Press.

Bresler, L. (2016). Interdisciplinary, intercultural, travels: Mapping a spectrum of research(er) experiences. In Burnard, P., Mackinlay, E. & Powell, K. (Eds.) *The international handbook in intercultural arts research* (pp. 321–332). London: Routledge.

Bresler, L. (2018). Aesthetic-based research as pedagogy: The interplay of knowing and unknowing towards expanded seeing. In Leavy, P. (ed.) *The handbook of arts-based research* (pp. 649–672). New York: Guilford.

Chödrön, P. (2013). *Living beautifully with uncertainty and change.* Boston: Shambhala.

Denzin, N. (1989). *Interactive interactionism.* Thousand Oaks: Sage.

Dewey, J. (1934). *Art as experience.* New York: Perigee Books.

Dreeben, R. (1967). *On what is learned in school.* London: Addison Wesley.

Eisner, E. (1979). *The educational imagination.* New York: Macmillan.

Eisner, E. (1982). *Cognition and curriculum: A basis for deciding what to teach.* New York: Longman.

Ellis, C. (2009). *Revision: Autoethnographic reflections on life and work.* Walnut Creek, CA: Left Coast Press.

Gendlin, E. (1981). *Focusing.* New York: Bantam Books.

Goleman, D., & Davidson, R. J. (2017). *Altered traits: Science reveals how meditation changes your mind, brain, and body.* New York: Avery.

Gombrich, E. (1956). *Art and illusion: A study in the psychology of pictorial representation.* Princeton, NJ: Princeton University Press.

Goodlad, J. I., Klein, M. F., & Tye, K. A. (1979). In J. I. Goodlad (Ed.) *Curriculum inquiry: The study of curriculum practice* (pp. 43–77). New York: McGraw-Hill.

Hearne, E. (2009). Bringing the story home: A journey with "Beauty and the Beast." In Hearne, B. & Trites, R. (Eds.) *A narrative compass: Stories that guide women's lives.* Urbana: University of Illinois Press.

Hearne, E. (2015). Ida Waters turns off the lights: The inside and outside of knowledge. In L. Bresler (Ed.), *Beyond methods: Lessons from the arts to qualitative research* (pp. 153–164). Lund, Sweden: Malmö Academy of Music. http://mhm.lu.se/sites/mhm.lu.se/files/perspectives_in_music10.pdf.

Hearne, E. (2017). "Your One Wild and Precious Life": A tale of divergent patterns in narrative and musical development. *Bulletin of the Council on Research in Music Education,* Fall 2016/Winter 2017, (210–211) 153–165.

Jackson, P. (1968). *Life in classrooms.* New York: Holt, Reinhart & Winston.

Lincoln, Y. & Guba, E. (1985). *Naturalistic inquiry.* Thousand Oaks, CA: Sage.

London, P. (2003). *Drawing closer to nature: Making art in dialogue with the natural world.* Boston: Shambala.

London, P. (2007). Concerning the spiritual in art education. In Bresler, L. (Ed.). (2007). *International handbook of research in arts education* (pp. 1479–1492). Dordrecht, the Netherlands: Springer, Conviction.

McConeghey, H. (2003). *Art and soul.* Putnam, CT: Spring.

Myerhoff, B. (1978). *Number our days.* New York: Simon & Schuster.

Pallaro, P. (Ed.) (1999). *Authentic movement*. London: Jessica Kingsley.

Perl, S. (2004). *Felt sense: Writing with the body*. Portsmouth, NH: Heinemann.

Peshkin, A. (1988). In search of subjectivity—One's own. *Educational Researcher, 17*(7), 17–21.

Polanyi, M. (1958). *Personal knowledge: Towards a post-critical philosophy*. Chicago, IL: University of Chicago Press.

Polanyi, M. (1966). *The tacit dimension*. Chicago, IL: University of Chicago Press.

Schwab, J. (1969). *The practical: A language for curriculum*. Washington D.C.: National Educational Association, Center for the Study of Instruction.

Spindler, G. (Ed.) (1982). *Doing the ethnography of schooling*. Stanford, CA: Stanford University Press.

Stake, R. (1967). The countenance of educational evaluation. *Teachers College Record, 68*(7): 523–540.

Stake, R., Bresler, L. & Mabry, L. (1991). *Custom and cherishing: The arts in elementary schools*. Urbana, IL: University of Illinois, Council for Research in Music Education.

Suzuki, S. (1970). *Zen mind, beginner's mind*. (Ed. T. Dixon) New York: Weatherhill.

Walker, D. (1989). *Fundamentals of curriculum*. San Diego: Harcourt Brace Jovanovich.

Wasser, J., & Bresler, L. (1996). Working in the interpretive zone: Conceptualizing collaboration in qualitative research teams. *Educational Researcher, 25*(5), 5–15.

7

VIRTUOUS INQUIRY, REFUSAL, AND CYNICAL WORK

Aaron M. Kuntz

It seems remarkably easy to become desensitized to the news of our day. That is, there remains a quick and determining reflex to shut down the entirely appropriate affect of outrage and shock given the inundation of daily transgressions against our collective sense of decency and shared ethical claims of right and wrong. In my experience, this desensitization dulls the otherwise sharp determination of moral outrage into a blunted sense of resignation and, well, political depression. Outrage and resistance are exhausting endeavors, after all; pulling back can generate a reprieve from the expended energy of engagement. Of course, as many of my colleagues point out, there is a degree of privilege that extends from those afforded the time and space to "check out"—to pause moral indignation for a moment in the hopes of achieving some semblance of quiet self-care. Simply put, not everyone can take the time to invoke such breathing room; the incessant assault of injustice continues unabated, felt on some bodies and relations more deliberately and consistently than others. This chapter is about outrage and refusal as a generative ontological formation, one enacted through truth-telling and truth-making. It is also about virtuous inquiry as an ethical imperative to create the conditions necessary for material change. In this sense, this chapter is about necessary claims on justice, right and wrong.

Normalizing Injustice

In much of academic discourse desensitization is often read as an implicit effect of normalizing processes: witness some indignity or affront to social justice enough and it becomes commonplace. Thus it is that some respond to such circumstance with a determination to intervene within processes of normalization—refusing the tendency to let some outrageous act cast a shadow over what previously had motivated a collective energy of refusal. In this instance the otherwise inevitable relation of desensitization with normalization is challenged through critique, and alternative effects become newly possible.

Despite these efforts, my sense is that processes of normalization depend, in large part, on a mistaken interpretive focus on acts extracted from the logics, assumptions, and conditions that make them possible in the first place.[1] That is, if it is the isolated act that remains the sole focus of one's critique, well, all it takes is another, more egregious act to make the former seem comparatively less important than the latter. In this instance, normalization extends as an inevitable effect of the many hyperactive injustices that plague our contemporary moment. It is in this way that critical inquiry perhaps fails its hopeful goal: specific actions may change but the values and assumptions that inform them do not; repetition wins out over difference and inquiry merely documents inertia.

In counter form, because repeated acts are often bound by a similarly enacted logic and entanglement of material conditions, their relation remains the productive focus of our critical inquiry. Indeed, engaged resistive inquiry can bring such relations to light in useful ways, manifesting the conditions necessary to enact material alternatives to the governing status quo. In a similar vein, Michel Foucault (1991) notes that his

> target of analysis wasn't "institutions," "theories" or "ideology," but *practices*—with the aim of grasping the conditions which make these acceptable at a given moment; the hypothesis being that these types of practice … possess up to a point their own specific regularities, logic, strategy, self-evidence and "reason."
>
> *(p. 75; original emphasis)*

Far from employing extractive logic, Foucault's target of *practices* reveals an immanent form of critique situated within a "given moment," refusing the isolationist tendencies of extracted analyses on the one hand and the overly general expression of totalizing theory on the other (both aiding the reductive cause of normalization).[2] To focus simply on the act is to miss the dynamic complexities of its enactment. To generalize about all-encompassing theories is to miss the immanently material manifestations of our reason. Perhaps more importantly, such hollow employments of critique also miss potential entry points for productive change; more must be done.

Thus, it is in this chapter that I argue for a collectively engaged and relationally enacted *virtuous inquiry* built on practices of radical refusal, truth-telling, and an ethical insistence that we might (must) become otherwise. This argument extends, I believe, from a parallel claim that we must not become desensitized to logics of injustice and inequity, no matter how often invoked to justify ethically abhorrent practices. In short, our inquiry work must actively refuse to normalize injustice. Further, we cannot pretend that practices exist outside of, or come after, the logics that inform them. Of equal importance, we must not allow concerns regarding normalization to overwhelm our own espousal of normative claims regarding what is good or just. That is, we must not allow our critiques of normalization to shortchange our own ability to enact normative claims as part of our resistive practices of inquiry. Far too often a collective hesitancy to invoke normative claims results in a type of frozen relativism wherein each and every position or relation is granted equal ethical status and informed change stalls. Given our contemporary moment, our inquiry work must do more than pretend at being neutrally descriptive or materially disengaged. Inertia (in both senses of the term—both doing nothing and maintaining in our present state) is not an option. Indeed, we might instead invoke immanently normative claims without, at the same time, reinscribing tendencies of normalization.

I do realize that foregrounding notions of virtue and tasking critical inquirers with making overtly normative claims perhaps invites charges of indulging humanistic and structuralist assumptions regarding inquiry. However, it is my hope that layering such terms with a relational materialism makes possible enactments of a productively resistive form of inquiry—inquiry practices that extend from a determination to foreground issues of justice within a contemporary moment that, in the collective, seems anything but just and far from fair. Indeed, it is my sense that we have an ethical obligation to refuse the exploitative relations and logics that seem to constitute our immediate moment even as we must use our inquiry to invoke a future unknown, oriented by a stance of immanent ethical creation. As such, engaging in virtuous inquiry might be a useful means to disrupt the otherwise inevitable march towards residual desensitization and normalization of morally abhorrent practices that threaten to overwhelm our collective consciousness (the collective affect of apathy that once was used as a descriptor for some youthful disengagement but has quickly spread to relations of all types).

Organizationally, I begin the next section by situating critique as a type of virtuous activity, one that aligns well with what I term *relational materialism* (Kuntz, 2019), a philosophical orientation that informs much of our contemporary materialist inquiry. This notion of critique articulates through an expression of immanent normativity, an of-the-moment force of the materially just and good. With such notions in mind, I then turn to a type of cynical engagement that might usefully inform our inquiry practices through an affect of disgust and abhorrence of the exploitative status quo. Finally, I conclude with the assertion that our times call for a type of virtuous inquiry, one imbued with the force of refusal and animated through cynical truth telling.

Virtuous Critique

To begin, it seems important to acknowledge the push–pull of a socio-political history of invoking liberal values (of agency, subjecthood, inherent rights, etc.) and developing relational orientations that are critical of the closure involved in such a position, emphasizing instead the intra-connectedness of a monist world-view. Because generationally learned habits of interpretation and practice are not so easily overcome, I find the liberal subject, for example, never fully absent from discussions of values and ethics in inquiry work. This remains the case even when those involved lay claim to a relational approach of "new material-ism," say, or "posthumanism." The specter of liberalism is certainly hard to shake. This plays out in rather interesting ways within the field of inquiry as many scholars seek to challenge the existing status quo in favor of ... well, something else. This, of course, is the very real struggle of *thinking* or *difference-making* within a normalizing culture bent on *repetition*.[3] And yet, given the ubiquity of brazen overt injustice in our world, these are not simply theoretical debates. Confecting ways to *make a difference* matters. Specific to inquiry prac-tices, such work begins to articulate through practices of engaged critique.

Somewhat in response to such tensions, Claire Colebrook (2012) provoca-tively pushes through the implicit liberalism of contemporary theory to encour-age an immanent and relational orientation to the construction of problems:

> Surely now is the time not to ask how "we" decide to maintain who "we" are but whether there might be questions, powers, problems that are not of our own choosing that affect us not as doers or performers but as barely adequate witnesses.
>
> *(p. 96)*

As more-than-liberal-subjects, "we" are not afforded the direct agency to choose select problems or relations arrayed around us—these are immanent contexts that extend beyond our individual capacity. "We" are not the origina-tors of selectively prescribed truths, nor may "we" lay claim to some innate ability to wholly define the problems at hand. And yet, despite this, we are perhaps bound by an ethical requirement to engage as "barely adequate wit-nesses" (yet witnesses nonetheless). Decentering a sense of agency does not release our ethical responsibility to those effects that extend from our relations, entangled through they may be. As Rosi Braidotti (2011; 2013) so often pro-claims, "we are all in this together." Perhaps it is the work of critique to pro-ductively engage an understanding of the *we*, *this*, and *together* of such a sentiment. This, to my mind, draws forth the ethical work of critique, making way for a type of virtuous inquiry within the relationally material world. To do so, we perhaps need to re-imbue the very notion of *critique* with an ethical charge, one that brings with it claims on the good and just.

In her examination of Foucauldian notions of *critique*, Judith Butler (2015) argues that such critical work deliberately contributes to *normative theory*, or, considerations for what is right or wrong, just or unjust. From this perspective, enactments of *critique* necessarily refuse the status quo (they are not simply descriptive) even as they are oriented by a belief that we might productively be other than we are—or, to use the terminology of the day, that we might *become* differently than we currently are (critique generates the potential of what has yet to be). Foucault layers this notion of critique with an aesthetic consideration of value, intersecting both ethical and political concerns within his examination of practices of virtue. That in his later writing and lectures Foucault engages with virtuous practice among the Ancient Greeks perhaps points to the degree to which contemporary critical practices contain only an echo of such normative claims. Instead, practices of critique among methodologists seem to have oriented away from questions of the good and just, fixating instead on disappointingly procedural concerns. Might we come to enact a sense of the good through practices of virtuous inquiry?

As Levy (2004) notes, once critical practices have shifted over time from considerations of virtue to a procedural ethics that evaluates acts solely "as contrary to or in conformity with the law" (p. 25). In this sense, inscribed law becomes the square against which daily practices are understood as ethical and valued (or not). What was once a consideration of virtue has become a methodological question of measurement: did you break the law? If so, to what extent? What was your degree of alignment with the law? (Within conventional research practices, the governing law remains validity or generalizability.) In this way, critical practices that once pointed towards unknown potential—considerations of what might yet become—shift towards enforcing a governance of what is—concern for reinscribing the standardized norm. Procedurized inquiry work thus takes on normalizing characteristics, losing critical potential.

For Foucault, being governed according to normalized domains comes with particular costs (to freedom, to democracy, to subjectivity) and thus critical practices might be reframed as ethical practices—politically engaged ways of becoming other than we currently are (or are coerced into being). Far from measuring degrees of normalized alignment, immanent critical work refuses to accept the costs of acting and remaining as we always have been, of accepting existence as governed subjects; critique newly born through insubordination. This, perhaps, is the beginning of a notion of virtuous inquiry as critical work, a form of resistance that engages an aesthetics of existence through refusal ("the art of not being governed like that and at that cost" as Foucault [1997, p. 27] termed it). Indeed, from this perspective, virtue articulates as a type of resistive critique mobilized through the force of enacted values. This causes Butler (2015) to write, "virtue is not only a way of complying with or conforming with preestablished norms. It is, more radically, a critical relation to those norms" (p. 313). Radical virtue takes on the force of disruption, challenging the regulation and assumed order of normalizing processes. Virtue thus orients as a critical relation to those practices through which normalization occurs.

Part of the role of virtuous critique would seem to necessarily include an interrogation of how select practices are induced to manage historically defined problems; that is, problems already containing solutions. This presents, of course, a question for inquiry—how do we generate creative inquiry practices that break the binds of normalizing inertia—problems without normalized solutions? Butler (2015) points to a similar quandary: "certain kinds of practices which are designed to handle certain kinds of problems produce, over time, a settled domain of ontology as their consequence, and this ontological domain, in turn, constrains our understanding of what is possible" (p. 4)—the closure and immobility of ontological inertia. Thus, part of the problem that plagues inquiry projects of resistance is that we encounter those habitual ways in which we have come to reenact "a settled domain of ontology" through research practices that locate prefabricated problems without making space for an unknown future in excess of normalized ontologies. We cannot learn to live otherwise amidst such circumstance (and live otherwise we must, such is our world today). To address this methodological bind, virtuous work engages in practices that challenge existing circumstance to produce previously unthought ways of living, yet-to-be defined ontological existences. Indeed, as Butler succinctly states, "liberty emerges at the limits of what one can know" (p. 8). New ontological possibilities manifest at the limits of epistemological practice; the threshold of knowing into becoming, encountered through virtuous inquiry.

On Normativity

To build towards a dynamic notion of virtuous inquiry I turn next to an alternative engagement with normativity—one that resists the essentialist tendencies of closure that plague conventional engagements with the term. Yet this is not such an easy task when some inquiry orientations refuse overtly normative claims. This results, as Colebrook (2012) critically observes, in a contemporary "war on normativity" (p. 85) that manifests in a distorted type of "ultra-humanism" often mischaracterized as a theoretical positioning of post-humanism: "we no longer believe in the privileged distinction of privileged white 'man', for *everything that lives* is an agent—subjected to the one norm of unity, community, communication, reciprocity and ecology: deep down we are all human" (p. 86; original emphasis). Ironically, through attempts to critique the traditional assumptions of liberalism, this ultra-human post-humanism extends from an unacknowledged privileging of liberalism itself. Most often, this manifests through applying liberal claims on the agential towards traditionally "non-human" entities or relations—everything is an *agent* and, because, as noted earlier, liberalism never fully disappears, the term retains its human connotations. Invoking the liberal (human) subject, even implicitly, assumes normative claims that remain hidden, veiled by the supposed displacement of humanism. Yet, if we get rid of the human agent, if we put liberalism under erasure, how are we to make normative claims (and we must, such is our responsibility of becoming in a world)? This remains a pressing question for inquiry projects bent on effecting material change.

Somewhat in response, Braidotti and Pisters (2012) offer a Deleuzian sense of "nomadic normativity" that foregrounds differential value creation and avoids "the free fall into relativism and nihilism" that often characterizes caricatures of anti-foundational philosophy (p. 1). Unlike conventional applications of morality that interpret acts against an external and timeless series of assumptions, "nomadic normativity" generates through "emerging, transversal collective affirmative values" (p. 2). As a result, normative claims extend through "immanent ethical principles of creation of values that concern not so much what ought to be, but rather what might be" (p. 2). Immanent relations, practices, and conditions are imbued with potential—*what might be*—and their open-ended entanglement articulates values that refuse the finality of closure.

To return to a focus on virtue, immanent affirmative values guide this practice towards what has yet to come to pass, an in-the-moment, process-based value system that begins with the recognition that all relations *might be* something more; virtuous potential to be otherwise. In this way, virtuous practice takes on the characteristics of immanent critique born of, and impacting, the relationally immediate now. As Braidotti and Pisters (2012) eloquently write, "normativity has become dynamic and creative, transforming reality always according to hidden intensities which ... present the ethical norm of finding new ways of how we might inhabit the earth" (p. 8). Immanent normativity takes on productive characteristics, generating newly configured practices and relations as an "ethical norm"—a valued way of becoming with the world. We are all in this together—*and* we might become differently, together. This is necessarily hopeful work.

This emphasis on virtue and immanent normativity perhaps also shifts the goal of inquiry away from attempts to generate universal truths or measures and towards a cartographic engagement of "the conditions under which something new is produced" (Braidotti and Pisters 2012, p. 6). That is, perhaps inquiry is now charged with mapping potential—the virtuous activity of generating relations yet to come to pass. As Kristensen (2012) puts it, "normativity has to be invented. It is this invention of new ground that makes it possible for a different way of thinking that can open new forms of action and belief" (p. 12). Perhaps inquiry, when imbued with virtuous force, can become a practice of ethically oriented difference making. To do so, inquirers would do well to engage a dynamic sense of truth and truth-telling, two important elements necessary to make normative claims through our work.

On Truth as Productive Refusal

Critical theorists and philosophers alike often seem hesitant to engage with notions of truth for any host of legitimate reasons: truths are essentialist products of modernity; truths are ahistorical, pretending to remain outside the impact of time or space; etc. Given the "material turn" that has developed of late, perhaps questions of truth are seen as, well, immaterial. Indeed, I came to examine questions regarding the conditions of truth-telling because I had become irritable, ticked even, at the hesitancy of my poststructural friends to entertain notions of truth—they would, it seemed to me, engage in exercises of linguistic gymnastics to avoid even mentioning the term (now, of course, everyone is talking about vacuous truths in the political arena—the amoral environment of post-truths). This hesitancy ran somewhat counter to my training in philosophy at a small Catholic college as an undergrad yet aligned well with my master's work in gender and queer theory, as well as with my doctoral training in neo-Marxism and materialism. So it was that I grumpily came to Foucault's explication of *parrhesia*, or truth telling—and later lectures on virtue. So here I am, dealing with truth (or, its telling anyway) and linking it to virtue—that other pesky term that points to the future not to say what one will do (that is prescription) but what one perhaps might do; an open-ended potential of what has yet to be.

Indeed, the very notion of truth indicates an excess of the contemporary moment—something unbound by the normalizing tendencies of our institutions. As Foucault (2011) asserts, "there is no establishment of the truth without an essential position of otherness: the truth is never the same; there can be truth only in the form of the other world and the other life" (p. 340). Truth generates as the creation of difference; the possibility that is "the other world and the other life." This living truth proffers a freedom not otherwise found: "a different life and a different world [that] have a real existence" (Prozorov, 2017, p. 820). The "real existence" of truth points to its material effects; truth as more than ideation.

Given this generative aspect of truth, perhaps a reorientation is necessary towards inquiry practices of truth-telling and those relational effects generated through such activity. Why is it that we inquire anyway? And, what types of future possibilities do we strive to manifest through our inquiry work? After all, much of our inquiry work is said to tell some element of truth, however constructed. Further (and perhaps most importantly), it does not seem wise to cede the ground on all things related to truth. Any retreat from the question of truth makes way, I believe, for uncomfortable claims of relativism as well as the destructive mechanisms of fascism (the practices of which are all too clear in my country's government in this present moment—"post-truth" and "fake news" as rhetorical means to distance destructive actions from ethical considerations and consequences). In response, perhaps the onus is on critical inquirers to more forcefully truth-tell and to do so with the deliberate intention of making newly possible resistive effects.

In short, I want us to take truth seriously. To do so, we need to disengage from mapping truth solely in regards to its correspondence to some external, empirical reality. This is a limited association of truth as reflecting something that came before its utterance. That is, we might productively situate truth as a practice—a telling or making—that generates material effects through its very production. Truth-telling makes possible a series of relations and actions not possible before its enactment. As a form of truth-telling, virtuous inquiry comes to articulate truths of the moment, an immanent critique that generates a future unknown.

In similar fashion, Kristensen (2012) situates truth as a type of critique: "the effort of creating new forms of problems"—a creative practice that challenges the present terrain to make way for alternative relations or contexts (p. 18). Truth, then, is not about identified solutions to some preexisting problem. Truth has a hand in making problems—a creative alternative to the dominant status quo. Through this understanding of problem-creation, Kristensen posits as the goal of philosophy to "develop concepts to apprehend something that does not yet exist but is about to come into existence" (p. 18). A similar goal might be said of the work of inquiry as we truth-tell on the precipice of change, motivated by the force of virtue. Of course, these notions of critique, truth, and truth-telling are not entirely new. As such, we might learn to enact virtuous critique as a type of *parrhesia*, or truth-telling, a forceful ontological orientation with philosophical roots in Ancient Greece.

On *Parrhesia* (as Virtuous Critique)

As a gloss, within Ancient Greek society, *parrhesia* existed under three main conditions: citizenship, responsibility, and risk.[4] In order to engage in *parrhesia*, one must be recognized as a citizen (and thereby able to operate in the public sphere). With such status, one has the ethical responsibility to speak truths to the public as they came to be known; citizenship thus brings with it the challenge of engagement. Thus, as a function of their citizenship, *parrhesiasts* make no attempt to occlude or otherwise hide the truths they understand and live. Such truths cannot replicate what is already known (else they would exist as technical knowledge, not truth) but instead point to possibilities for becoming otherwise (virtuous existence that has yet to fully form). As a living practice, the telling of such truths disrupts the legitimated order, calling into question the very circumstances that grant the truth-teller citizenship. In this way, *parrhesiasts* seek to transform the relations of which they are a part through the virtue of truth-telling; a type of living critique. Importantly, the *parrhesiast* never fully knows—never could fully know—the outcome of truth-telling and, as such, each *parrhesiastic* activity carries with it some degree of (productive) risk.

Within democratic political circumstances, the *parrhesiast* is made vulnerable to the public, risking the ability to control another's response to the truths told. There is thus a precarity to *parrhesia*, a vulnerability that extends from a productive release of containment: neither truth-telling nor the collective response to such telling can be prescribed or anticipated. Butler's (2015) determination to "think vulnerability and agency together" (p. 139) perhaps aligns well with the vulnerable risk that comes from truth-telling for material change, particularly as a public act of refusal. For in articulating a truth, a challenge to the governing norm, one becomes vulnerable to the discipline afforded the governing system even as one contributes to a shared enactment of distributive agency; the building effects of collective refusal. As such, there exists a shared vulnerability within truth-telling, a responsibility to speak with as opposed to only speaking to. Inquirers and philosophers alike can learn from such a resistive orientation.

Importantly, Foucault situates *parrhesia* as unbound by the confines of law—an excess to jurisdiction. *Parrhesia* thus avoids the technical questions of "is it lawful?" or "what is the measure of its truth?" (those questions of convention referenced earlier) in order to ask more engaged questions regarding moral conduct given the emergent contexts of the world (questions of virtuous becoming) (Foucault, 2001). *Parrhesia*, then, is never complete, never fully determined, but extends as an orientation to the world that begins with determinations for change. Thus it is that technical questions give way to ethical considerations of living practices.

Given this, Michael Peters (2003) situates the "problem of *parrhesia*" as the question of who has the right, duty, and courage to speak truth (p. 214). The short answer to such a problem, of course, is anyone who enjoys the privilege of citizenship. However, I wonder if we might push this question a bit further to ask, *who has the right, duty, and courage to confect the conditions such that truths might be told?* Perhaps this is a productive role for inquiry given the materially disheartening times in which we live. And, given what I have written thus far, perhaps enactments of virtuous inquiry are the means through which such conditions might be realized. Specific to my arguments in this chapter, *parrhesia* articulates as a cultivation of self-in-relation (like virtue); a becoming subject. This open-ended orientation foregrounds potential—what might be—through the normative work of critique. With *parrhesia* as a guide, virtuous inquiry articulates as enactments of normative claims for productive change—the immanent challenge of truth-telling to manifest a potential unknown. Such work perhaps begins through a resistive orientation to the world as-is.

As Seitz (2016) points out, *parrhesia* extends as a practice of disagreement with contemporary norms, articulating as a "dissensual conception of truth" (p. 5). Agreed upon truths are refuted by the *parrhesist*, making way for immanent truths in the making. In this way, the truth-teller enacts a living freedom, one of "not being entirely governed by pre-existing rules of discourse" (p. 5). *Parrhesia* extends then as "an immanent provocation to all normative discourse" that makes impossible the normalizing tendencies of consensus (p. 10). Truth-telling remains disruptive, confronting normative discourse such that it is not easily repeated (or repeatable). Our inquiry work might well follow suit.

One example of truth-telling as a productive means of refusal extends from the practices of the Cynics as they employed their public visibility to invoke outrage and shock at the injustices of the governing status quo in Ancient

Greece. Given the egregious violations of basic rights that seem the political norm today, I find the outraged and passionate stance for change that characterizes cynical truth-telling useful, even productive. Cynical *parrhesia* differentiates itself from other forms of truth-telling through insisting on a collapse of the previously bifurcated elements of life and truth; the one no longer coming before the other, they merge such that "the body gives form to truth and truth gives form to body" (Lemm, quoted in Prozorov, 2017, p. 807). Through this intersection, life/truth lose the disproportionate relation previously claimed through other forms of *parrhesia* (wherein life was dominated by truth; life understood and lived through its relation to external and atemporal truth). For the Cynics, to separate truth and life is to distort their relation; confecting a distinction where there is only entangled intra-relation (Prozorov, 2017). Through cynicism, life and truth are simultaneously enacted.

As part of their material challenge to the status quo, Cynics strove to provoke affective responses of disgust and revulsion to conventional practices of living. As Foucault (2011) describes it, cynical practices resembled a "broken mirror" that reflects both what one is and what one "would like to be" (p. 232). Witnessing the doubled image of cynicism's broken mirror—the fold of the immediate now with a future potential—results in a "grimace, a violent, ugly, unsightly deformation" that is unrecognizable to the conventional eye (p. 232). Thus, the Cynics transform all of their conventional relations through their practices of radical critique and refusal. Through the broken mirror, cynical *parrhesia* articulates as strategies of "getting people to condemn, reject, despise, and insult the very manifestation of what they accept, or claim to accept at the level of principles" (Foucault, 2011, p. 234). Through this work, conventional practices take on grotesque formation, no longer easily enacted or recognized. Indeed, such normalized activities take on violent characteristics—to enact them is to invoke the violence of normalized repetition. Lastly, cynical truth-tellers refuse the ordered space of the periphery: "In their insistence to cause trouble, the Cynics demand to be seen" (Kuntz, 2019, p. 139).

In these ways, cynical *parrhesia* draws force through acts of disobedience, a challenge to the norms and conventions that dominate the immediate moment (Prozorov, 2017). Thus, we would do well to consider how such an orientation productively alters practices of inquiry. Inquiry informed by cynical *parrhesia* makes possible newly formed challenges to what is, in the hopes that different relations might productively emerge out of such work. Inquiry is thus begun as an ethical practice of disobedience. It is to make possible another life, another series of relations not previously possible.

Contemporarily, refusal or insubordination is never enacted in isolation—these are relational practices—*we are all in this together*. As such, cynical *parrhesia* makes possible a collective challenge to the governing status quo. Further still, in our contemporary moment cynical *parrhesia* might manifest as a type of virtuous inquiry, making possible the circumstances necessary for relational truth-telling to occur, bringing together as it does important elements of citizenship, ethical duty, outrage, and vulnerable risk.

Learning from cynical practices of *parrhesia*, we perhaps need inquiry work that effects a coming together—a collective assembly of refusal and resistance. As Butler (2015) notes, the open-ended orientation of virtue extends as a practice of desubjugation wherein a subject "risks its deformation as a subject, occupying [an] ontologically insecure position" (p. 12). There is risk to such virtuous practice, refusing the surety of stasis and repetition. Together, we *grimace*, the *unsightly deformation* of what can no longer be and thus must become otherwise. Virtuous inquiry requires that we become something not yet formed, monsters in the making.

The civil rights icon and Georgia congressman John Lewis often talks of "good trouble" (on Twitter, Lewis tends to conclude his tweets with "#goodtrouble"). As Lewis recently tweeted, "Sometimes you have to get in trouble—good trouble, necessary trouble—to make a way out of no way" (@repjohnlewis, 2018, March 14). Note here the creative force of Lewis' declaration: making "good trouble" produces "a way out of no way." Through this process, constraint and governance give way to potential; what is becomes what might yet be. Given our contemporary moment, perhaps inquiry could engage in similarly resistive work, driven by the collective force of outrage, truth, and a virtuous orientation to necessary change.

Through our inquiry work, we need more than truths spoken—we need confronting truth; tellings that challenge, intervene, and affect another towards change. We perhaps, given our moment, need cynical truths. In such circumstance, inquiry enacts changes, not simply describing the world but creating new problems and generative concepts to ethically transform our material relations towards something else. As I have noted elsewhere (Kuntz, 2019), "through virtuous inquiry we perhaps seek to short-circuit the exploitative relations of power" that maintain injustice in our shared world (p. 20). Similarly, Denzin and Giardina (2018) write, "inquiry is always a form of moral intervention in the service of liberation" (p. 12). Perhaps virtuous inquiry is a means to orient such intervention, to realize the potential of such liberation.

Notes

1 In my previous work (Kuntz, 2015) I termed this as operating from a methodological *logics of extraction*.
2 Of course, this remains part of Foucault's challenge—to develop a *history of the present* (no easy task, that).
3 The Deleuze-inspired terms are intentional here: *thinking* and *repetition* and *difference*.
4 For a more thorough treatment of *parrhesia* within inquiry, see my earlier work (Kuntz, 2015). For a sustained engagement of truth within materialist inquiry, see my most recent book (Kuntz, 2019).

References

Braidotti, Rosi. 2011. *Nomadic Theory: The portable Rosi Braidotti*. New York: Columbia University Press.

Braidotti, Rosi. 2013. *The Posthuman*. London: Polity Press.

Braidotti, Rosi and Pisters, Patricia. 2012. "Introduction." In *Revisiting Normativity with Deleuze*, edited by Rosi Braidotti and Patricia Pisters, 1–10. London: Bloomsbury.

Butler, Judith. 2015. *Notes Toward a Performative Theory of Assembly*. Cambridge, MA: Harvard University Press.

Colebrook, Claire. 2012. "Norm Wars." In *Revisiting Normativity with Deleuze*, edited by Rosi Braidotti and Patricia Pisters, 81–97. London: Bloomsbury.

Denzin, Norman and Giardina, Michael. 2018. "Introduction." In *Qualitative Inquiry in the Public Sphere*. Edited by N. Denzin and M. Giardina. New York: London. 1–14.

Foucault, Michel. 1991. "Questions of Method." In *The Foucault Effect: Studies in Governmentality*, edited by Graham Burchell, Colin Gordon, and Peter Miller, 73–86. Chicago, IL: University of Chicago Press.

Foucault, Michel. 1997. *The Politics of Truth*. Translated by Lysa Hochroth and C. Porter. Los Angeles, CA: Semiotext(e).

Foucault, Michel. 2001. *Fearless Speech*. Edited by Joseph Pearson. Los Angeles, CA: Semiotext(e).

Foucault, Michel. 2011. *The Courage of the Truth (The Government of Self and Others II): Lectures at the College De France 1983–1984*. Translated by G. Burchell. New York: Palgrave Macmillan.

Kristensen, Anders. 2012. "Thinking and Normativity in Deleuze's Philosophy." In *Revisiting Normativity with Deleuze*, edited by Rosi Braidotti and Patricia Pisters, 11–24. London: Bloomsbury.

Kuntz, Aaron. 2015. *The Responsible Methodologist: Inquiry, Truth-Telling, & Social Justice*. New York: Routledge.

Kuntz, Aaron. 2019. *Qualitative Inquiry, Cartography, and the Promise of Material Change*. London: Routledge.

Levy, Neil. 2004. "Foucault as Virtue Ethicist." *Foucault Studies* 1: 20–31.

Peters, Michael. 2003. "Truth-Telling as an Educational Practice of the Self: Foucault, Parrhesia and the Ethics of Subjectivity." *Oxford Review of Education* 29(2): 207–223.

Prozorov, Sergei. 2017. "Foucault's Affirmative Biopolitics: Cynic Parrhesia and the Biopower of the Powerless." *Political Theory* 45(6): 801–823.

Seitz, Sergei. 2016. "Truth beyond Consensus – Parrhesia, Dissent, and Subjectivation." *Epekeina* 7(1): 1–13.

St. Pierre, Elizabeth. 1997. "Circling the Text: Nomadic Writing Practices." *Qualitative Inquiry* 3(4):403–418.

8

THEORIZING FROM THE STREETS

De/colonizing, Contemplative, and Creative Approaches and Consideration of Quality in Arts-Based Qualitative Research

Kakali Bhattacharya

Over the past 15 years, I have persistently moved towards anti-colonial onto-epistemologies in education and beyond by utilizing de/colonizing approaches to thinking about and engaging in qualitative inquiry. Through this journey I have found myself in various contested and intersecting spaces of existing and new theoretical and methodological discourses and praxis. Sometimes I resisted these discourses. Other times I was complicit, and still at other times, I escaped from such discourses to spaces of pure absurdity, play, and nonsense, carving out varied possibilities. Often such spaces have been informed by contemplative, de/colonial, and creative approaches to qualitative inquiry. In this chapter, I will trace these paths and my negotiations of their possibilities and productive tensions.

A State of Confusion: Searching for Entry Points

There has been a discursive proliferation of post-qualitative methods privileging new materialism and posthuman approaches to qualitative inquiry. Central to this proliferation are ideas of decentering the human, radical interconnectivity between living beings and the non-living, and a refusal of methodology. Unfortunately, I have yet to find anything uniquely compelling to draw me to them. Allow me to elaborate. Through conversation with other minoritized scholars, it has become evident how much well-established communal knowledge and understanding are now part of posthuman and new materialist discourses without proper attribution or acknowledgement of, or access for, several minoritized communities. I hungered for those who adopted and promoted these discourses to interrogate their production of methodological complexities, their refusal of methodology or theory-driven inquiry. But such interrogation has been limited. At professional conferences, I witnessed a pervasive hierarchy that privileged high theory as sophisticated while dismissing all other theoretical approaches as crude, as if those who did not engage in high theory were simpletons, theoretical Neanderthals.

As a multiply minoritized woman of color, I have consistently been on the margins of these high theory discourses. In graduate school, I read the philosophical discourses of dead white French men and those who translated them, works that seemed dense and impenetrable. In one of my first response papers I tried to describe the distance I felt between my own sociocultural location and history and a particular author's work, and the alienation I experienced in response. My professor gave me a B with only one comment on the paper: "Lots of people read X and find it okay."

To me, this feedback ascribed a deficit narrative to me, which was represented as a result of my failure to make such dense, obscure discourse intelligible to myself. While this discourse prided itself on deconstructing grand narratives, its proponents used this same discourse as their own grand narrative. How are we to decenter dominant narratives if we continually create new centers that replicate the inherently dangerous power structures and their discursive material effects of oppression? Receiving such deficit-imposed feedback early in graduate school alerted me to the lack of institutional willingness to make space for me to speak from and about my location. I learned to engage in a performativity that portrayed a forced congruence with the readings, even though it was incongruent with my own ontoepistemic orientation.

It was not until later in my graduate training, when I read Linda Tuhiwai Smith's (1999/2012) *Decolonizing Methodologies*, that I learned about the colonizing nature of research, education, and higher education. It was at this point that I committed to unlearning what was recognized as the canon, the knowledge that was privileged in my discipline. For the past 15 years, I have been on this journey of unlearning.

In this chapter, I demonstrate this process of unlearning using the key idea of *theorizing from the streets*. Theorizing from the streets highlights the ways in which the unlearning of colonial and privileged discourses has unfolded in my scholarly work in qualitative inquiry engaging in knowledge-making expansively within and beyond academia. Please note that my work is not to be read in opposition to new materialistic and posthumanistic discourses (Barad, 2007; Braidotti, 2013). Instead, this work is additive, supplementing that which currently exists, creating a space for those whose lives and experiences do not resonate with the discourses currently privileged in qualitative inquiry. Rather than offering a comprehensive critique, I will highlight two key points of my incongruence with these privileged discourses that position me at the receiving end of ontoepistemic violence, with limited entry points to these discourses.

First, I cannot stop centering the human, because to do so is based on the presumption of a shared agreement about the status of "human." We know from embodied, minoritized perspectives that this simply is not true. Many communities lack the luxury of having their dignity and their humanity recognized, let alone centered. Therefore, decentering those who are already rendered dehumanized, invisible, and oppressed, further reinforces the inability of the subaltern to speak, to be heard, or to be seen (Spivak, 1993).

Additionally, the centering of human experience does not automatically imply a decentering of people from their ecological contexts. Indeed, in qualitative inquiry we offer framings and frames of our research. We recognize that there is both a foreground and a background to this framing, positive and negative spaces in the pictures we paint, and multiple points of view that allow us to focus and blur various parts of our inquiry. Decentering the human in her ecological context is unethical to me, especially for those who are always already decentered and dehumanized. That an ethnographer or a qualitative researcher should decide when humans matter and when they do not and in what ways, has been the reason why research in general, and qualitative research in particular, is colonizing and oppressive (Smith, 1999/2012).

Growing up in India, I learned early on about the radical interconnectivity among all things, without anyone specifically pointing this out to me. In Hinduism, our gods and goddesses are associated with their own pets, which are often birds, reptiles, insects, or other animals. Each god and goddess also has his or her own specific ritualistic preference about which types of grass, leaves, fruits, etc. are most appropriate for making offerings to them. As soon as I was aware of myself and the world around me, this type of interconnectivity centering and decentering humanized gods and goddesses, nature, and other living beings was so naturalized that it was as uninteresting as the weather, as it was fairly intuitively obvious. Without anyone instructing me, from the moment I became aware of my relationship with the world, I recognized a divine connection between me, a source energy, other sentient beings, and inanimate objects, comprising a collective consciousness. Consequently, I am not sure how to be excited about, or create entry points into the currently privileged discourses in qualitative research. My scholarly work has always taken into account such interconnectivity without being compelled to repeatedly point out in transparent ways how such moves were made.

Second, privileging the refusal of methodology, as many post-qualitative discourses promote (St. Pierre, 2013), seems at once a situated and established practice that is several decades old. For example, for my dissertation in 2005, I wanted to work with de/colonizing methodologies in the context of higher education, utilizing transnational feminism and creativity as a form of inquiry. I could not find any appropriate methodological options to guide me in regard to data collection, analysis, representation, or even trustworthiness. As a novice qualitative researcher who had to satisfy a faculty committee with their own preferences for certain structures within qualitative inquiry, I vacillated between the open-endedness of my preference for theory-driven inquiry and refusal of traditional methodology on the one hand, and my need to fulfill committee expectations to establish some familiar structure on the other.

Consequently, I offered the structure of a transnational feminist case study while deconstructing the assumption of holistic inquiry associated with case studies. I wrote an ethnodrama in a fluid format, comprising multiple, interconnected front- and back-stage plays, extending Goffman's idea of the performative self (Bhattacharya, 2009; Goffman, 1959). However, when I was expected to offer some themes at the end of writing the ethnodrama, I was stuck. I navigated through this stuck place using theory again. I offered thematic headings replete with contradictions and tensions, while the expository narrative that followed problematized fixed categories and depicted multiple entangled and messy contradictions and tensions.

While I privileged theory-driven inquiry, I also navigated the academic terrain in ways that supported completing my degree. Therefore, my theory-driven inquiry and tempered refusal of methodology were contingent on the context in which I had to exist and earn my degree. Earning my degree meant creating a presence for those who look like me while also de/colonizing established approaches to inquiry in higher education. It meant interrogating the discourses that promote certain ontoepistemic moves and turns, and the power relations and privileges of those who make such moves and turns.

Qualitative approaches that are incongruent to certain ontoepistemic turns should not be categorized as hierarchically inferior. For example, while there might be turns to hybridize ontology and epistemology, there are several minoritized communities that existed in such spaces of knowledge-making before the linguistic articulation. Their ways of inquiry did not require an ontoepistemic hybridization or turn, as it was the ways in which they already navigated their worlds. Thus, a refusal of methodology, emerging from certain ontoepistemologies, must be understood as a situated position, with its own limits and possibilities, that can open up multiple expansive terrains of inquiry, as evidenced in my arts-based work (Bhattacharya & Payne, 2016).

Additionally, when I worked with the Kansas State Department of Education to help them revise the accreditation system for schools throughout Kansas, we did not have the luxury of refusing methodology (Bhattacharya, 2016a). The politics and culture of this context required us to outline our methodologies, paths of and rationale for data collection, analysis, and representation. We were in dialogue with educational leaders, teachers, parents, businesses, higher education professionals, students, politicians, and policy makers.

Thus, in those contexts, while it would have been luxurious to explore what we might produce by refusing methodology, it was simply not feasible or productive. Therefore, we must recognize the privilege of positionalities/context that allow us to engage in inquiry without being restricted or restricting ourselves. We must simultaneously recognize the barriers others confront in negotiating the work they do in their own spaces, rife with their own challenges.

For example, certain disciplines are still quite post-positivist in terms of the kinds of qualitative research they are willing to accept. A researcher in that space who desires to expand the field might be unable to fully refuse methodology if she wishes to be published in her desired disciplinary journals.

Moreover, when such scholars occupy a minoritized position, they are often critiqued simply for working with their own communities, let alone doing so in the absence of any methodological engagement. This is not to say that a scholar in such a location cannot engage in boundary-expanding work. However, there are many positions to occupy within the continuum between post-positivism and refusal of methodology. Thus, any critique of the failure to refuse methodology must be accompanied by an examination of contextual possibilities and barriers.

Theorizing from the Streets: Which Moves Make my Work Matter?

In this section I take the reader through some of the critical moments in my journey of unlearning that I highlighted at the 2018 International Congress of Qualitative Inquiry. Using some of the slides I presented during the conference, I write about my unlearning and the possibilities and barriers I negotiated.

Since 2005, I have found myself repeatedly in positions that call for the unlearning of much of what I was taught and exposed to in my previous education. Having read *Decolonizing Methodologies* (Smith, 1999/2012), it was as if I could not return to the ways I formerly knew how to engage in qualitative inquiry. My search for *possibility models*, as in what has been made possible thus far, seemed a futile effort, as qualitative researchers are cautious about putting out work that could be read as prescriptive. Therefore, the paths I have forged in my unlearning have emerged through experimentation, creativity, criticality, and de/colonial and transnational feminist ontoepistemic orientations.

Most qualitative research is generated from countries located in the Global North. Most of the scholars who are celebrated in these spaces are white, whether they are writing traditional introductory textbooks or articles about deconstruction, new materialism, or posthumanism. Within these spaces, qualitative research is mostly celebrated through a form of whiteness, which (even if it is self-aware) cannot generate any knowledge that is completely devoid of whiteness. Most qualitative research faculty are also white, and they often teach using privileged texts in qualitative research. Thus, the training of qualitative researchers often lacks a critical self-interrogation of whiteness, which is vital for becoming expansive and de/colonial.

Imagine what kind of education we could create if we ceased to privilege a Global North sensibility in the teaching of qualitative research. What if we made space for the ontoepistemic direction of knowledge flow from east to west, or south to north? What if we were all willing to unlearn what we have held onto for so long?

Yet that is not the space within which qualitative research operates–especially critical qualitative research. Despite offering sophisticated theoretical critiques of established dominant, intersected discourses of race, class, gender, ableism, and sexuality, qualitative researchers nevertheless remain methodologically complicit with privileged texts and discourses in qualitative inquiry. In other words, a study that critiques the racist, colonial enterprise of schooling is likely nevertheless to cite the works of Creswell, Patton, and Maxwell in its methodology, as if such works are devoid of history and cultural orientation. If we accept that qualitative research is a colonizing enterprise, then we must actively construct new approaches to inquiry that are consistent with our ontoepistemic orientation and that honor the communities with whom we work. Without this shift, our studies will be culturally incongruent, and all of our critique and deconstruction will remain limited to theoretical readings of data.

In searching for ways to think about and conduct qualitative research in the liminal spaces of un/learning privileged approaches, I began to look for citational authority and knowledge making beyond academia. Academic gatekeeping is evident in the ways we evaluate the quality of a scholarly work based on its language, citations, and structure. Often these moves are not innocent. By continuing to privilege the already privileged in academia, they forward the notion that knowledge making that has worth and value exists only in these privileged academic spaces, and only under certain disciplinary gazes. Consequently, *theorizing from the streets* became a conscious effort to integrate information from outside the academic gaze, to expand what we consider to be legitimate sources and forms of knowledge.

The first move I made to expand academic spaces of knowledge making occurred in my dissertation, where I interrogated our posturing as interviewers who enter a research space. Using a song by Ludacris, entitled, *Stand Up*, I highlighted the *presumptive agency* of the researcher (Bhattacharya, 2009). The song demonstrated how a woman gets chosen by Ludacris to be in the VIP area of a club from the long line outside. Ludacris states, "When I move you, move, just like that," as he leans into a woman against the wall, and she responds by leaning back against the wall. I argued that as a researcher, we have the choice of participants and we engage in selecting our chosen ones, as Ludacris did. Armed with our techniques and methodologies, we posture towards our participants to be responsive to our moves, presuming and assigning ourselves more agency than the participants, thereby creating a submissive participant to our dominant agentic researcher. Figure 8.1 presents the figuration of such presumptive agency as a wall, with which a researcher might interact with a participant to *extract* information, limiting the participant to the scope of the study, and cornering her with probing questions to elicit explanations and clarify contradictions in her responses.

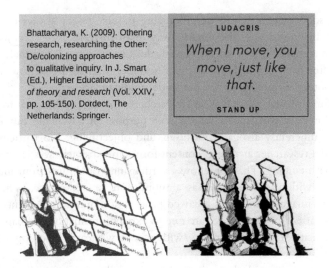

FIGURE 8.1 Presumptive agency of the researcher

Yet such limitations are merely projected by the researcher onto a participant as she is pushed up against some wall or bounded in restrictions of the research scope; a participant may not even perceive them as limitations and may thereby demonstrate her own agency. As researchers, then, we have to contend with meeting the participant where she is and *humbly* accepting what she is offering. Advocating for surrendering a will to know, I stated:

> The wall of presumptive agency demonstrates the building blocks of the epistemology and methodology that inform the strategies I incorporated partially in this study. Armed with open-ended questions and shared understanding, I might have the privilege of interrogating—asking for further explanation of answers. Since the wall of presumptive agency is constructed mostly through my researcher epistemologies it rests on unstable foundations, especially if the interviewee chooses to exercise her agency, rendering the wall invisible and moving to unanticipated and unimagined spaces. For the participant, the wall behind her does not exist in the same way it does for me. In light of this realization, I questioned the presumptive agency with which I designed this study. Eventually, this research emerged from the cracked space of the wall, within which I question the commensurability of methodology and de/colonizing theories.
>
> (Bhattacharya, 2009, p. 119)

This interrogation of my posturing as a researcher reflects both my own alignment with established practices and my need to unlearn some of those practices.

In doing so, I recognize that de/colonizing inquiry might always occur in a liminal, hybridized space, and might always exist in relation to colonizing discourses, materialities, and effects. From this space, knowledge making may never be free from coloniality. Yet as a researcher, I am aware of how my anticolonial ontoepistemologies burst out in de/colonizing desires.

The fundamental idea that as researchers we are authorized to obtain information from participants for the sole purpose of advancing our research agendas is inherently arrogant, intrusive, and objectifying. No participant has ever owed a researcher anything, consent form (an extremely westernized practice) notwithstanding. That somehow a participant would mirror our interrogative stance with their response, as a submission to our stance, is not only offensive to imagine but also is a flawed process of inquiry. Participants are not data repositories from whom we are free to extract information. Perhaps instead they are agents of wisdom through/with whom we learn, based on whatever they are willing to offer.

Such a stance demands that we surrender our will to know (a refusal of methodology) when we enter a research space. It requires us to recognize the sacred nature of conducting inquiry as we relate not only to our topic of inquiry, but also to the interconnected nature of the participants' lives from which they make their offerings. We must relinquish our expectation that we are entitled to information simply because a consent form was signed. Consent forms are products of western practice that falsely imagine the researcher's relationship with the participants as fixed. In truth, the researcher and participant relationship, like all relationships, is fluid, shifting, messy, and entangled (Bhattacharya, 2007).

Interrogating one's own de/coloniality requires a deep, introspective journey, akin to meeting one's fear and confronting the darkness within. As once-colonized subjects we might crave the master's approval or desire to be more like the master, representing the enslavement of our spirits. Indeed, Gloria Anzaldúa identified this need to gain our master's approbation with our complicity in our oppression when she said:

> On that day I say, "Yes all you people wound us when you reject us. Rejection strips us of self-worth; our vulnerability exposes us to shame. It is our innate identity you find wanting. We are ashamed that we need your good opinion, that we need your acceptance. We can no longer camouflage our needs, can no longer let defenses and fences sprout around us. We can no longer withdraw. To rage and look upon you with contempt is to rage and be contemptuous of ourselves. We can no longer blame you, nor disown the white parts, the male parts, the pathological parts, the queer parts, these vulnerable parts. Here we are weaponless with open arms, with only our magic. Let's try it our way, the *mestiza* way, the Chicana way, the woman way."
>
> (Anzaldúa, 1987/1999, p. 110)

In this vulnerable confession, Anzaldúa offers us a sense of radical interconnectivity between the various fragmented, wounded parts within us and the world at large. This radical interconnectivity is part of a messy de/colonizing move that emerges from the liminal, hybridized mestiza consciousness. Those of us with a history of colonization (whether as subjects, beneficiaries, or both) need to contemplate the ways in which our positionalities are shaped as a consequence of colonization. We need to engage in what Jung (2014) calls *shadow work*, exploring the manifestations of a darker archetypical version of self. According to Anzaldúa (2015), shadow can also be a consequence of trauma-based events in our lives that incentivize us to engage in repression, instead of gaining knowledge when we face and engage with the shadow. Shadow, then, is both personal and collective, reflecting ourselves, our families, our communities, our nations, and our world (Bhattacharya, 2018b).

Maheshweta Devi, a Bengali author, demonstrated shadow work in her novel *Shishu* (*Children*), where she described the "literal and figurative crippling of people in the lower classes and castes—the tribal people in India" (Bhattacharya, 2005, p. 52). In post-independence India, those who gain privilege are:

> those that can adopt the colonized views as their own. Thus, Mr. Singh, a character in *Shishu*, though well-intentioned, displays his colonizing attitudes towards a group of tribal people who were forced to live in the forests due to their rebellious demands for their lands and rights, and their need to practice living according to their cultural perspectives. Mr. Singh encounters these tribal people in the forest and thinks that they are mysterious, superstitious, uncivilized, and backward—those people that were once thought to be the indigenous pride of the country. At the chilling climax of the tale, we are brought face-to-face with these "children" who thrust their starved bodies towards Mr. Singh, forcing him to recognize that they are not children at all, but adult citizens, stunted by free India.
>
> *(Bhattacharya, 2005, p. 52)*

My use of an excerpt from a Bengali novel in my dissertation is anti-colonial, contemplative, and creative, comprising a knowledge-making source that allows me to theorize from the streets. First, the novel itself is written in Bengali, even though I provide an English translation. My initial access to the novel was in another language; thus, my embodiment of its knowledge making exceeds an anglocentric sensibility. Yet, I am complicit in producing anglocentric sensibilities when I translate, frame, explain, or elaborate on my positions to an anglocentric academic audience in the Global North.

Second, the novel embodies Maheshweta Devi's deep contemplation of our states of existence, our desire to be free while being forced to live under colonizing conditions. The use of a literary novel at the time when I was writing my dissertation, without any precedents in my immediate discipline, program, or

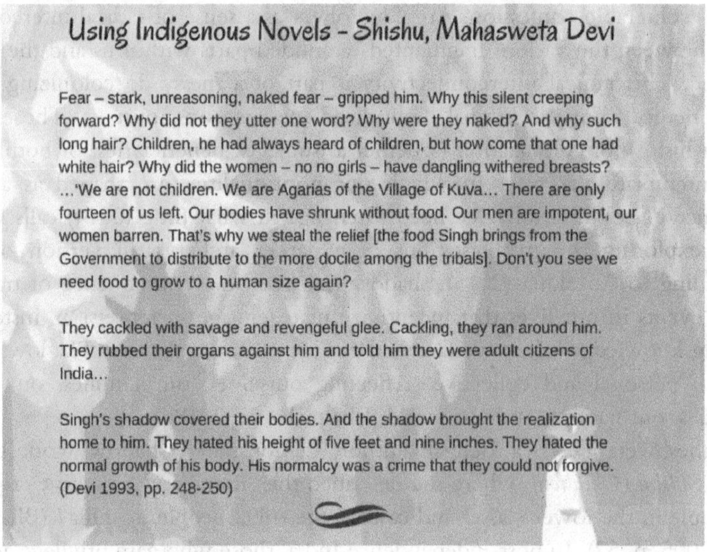

Using Indigenous Novels – Shishu, Mahasweta Devi

Fear – stark, unreasoning, naked fear – gripped him. Why this silent creeping forward? Why did not they utter one word? Why were they naked? And why such long hair? Children, he had always heard of children, but how come that one had white hair? Why did the women – no no girls – have dangling withered breasts? ...'We are not children. We are Agarias of the Village of Kuva... There are only fourteen of us left. Our bodies have shrunk without food. Our men are impotent, our women barren. That's why we steal the relief [the food Singh brings from the Government to distribute to the more docile among the tribals]. Don't you see we need food to grow to a human size again?

They cackled with savage and revengeful glee. Cackling, they ran around him. They rubbed their organs against him and told him they were adult citizens of India...

Singh's shadow covered their bodies. And the shadow brought the realization home to him. They hated his height of five feet and nine inches. They hated the normal growth of his body. His normalcy was a crime that they could not forgive. (Devi 1993, pp. 248-250)

FIGURE 8.2   Excerpt from Shishu

college, represented a move to use creativity to unlearn the colonizing influence on qualitative research. The story demonstrates a layered complexity from the perspective of both Mr. Singh, an accommodating colonizing subject, and the tribal people, who suffer from that which benefits Mr. Singh. Mr. Singh's gaze is colonizing as he views the tribal people as backwards and uncivilized, without recognizing that his complicity with and access to certain benefits cultivates the impoverished, dehumanizing conditions of the tribal people he encounters.

Theorizing from the streets thus constituted an ontoepistemic, methodological, contemplative, and aesthetic move/turn from that which was privileged, traditional, and/or deconstructive. I have defined and elaborated on contemplative practices elsewhere (Bhattacharya, In PressA). Here I will simply re-state that, "broadly speaking, contemplative practices refer to a willingness to travel within, to be mindful of one's actions, words, thoughts, to create deep awareness of self in relation to the world, and in relation with various aspects of self" (Bhattacharya, In PressA, p. 8). In my dissertation, I could not simply cite Foucault, Derrida, or other French philosophers to engage with the narratives of participants who were transnational students from India in their first year of graduate studies in the U.S., especially when I hungered for a more culturally grounded framing. Therefore, I had to look for other locations and sources to legitimize such an approach and expand the academic space in which I was working.

Ten years after completing my dissertation, I published an article (Bhattacharya, 2016b) about being a transnational academic of color in higher education in the U.S. In this article, I constructed a metaphor inspired by the movie *Life of*

Pi, in which Pi was an Indian teenager who was stranded after a shipwreck on a boat with a royal Bengal tiger named Richard Parker. I used a hybridized form of magical realism, fiction, and creative non-fiction to articulate the tensions of navigating academia from my location. I described my de/colonial negotiations in academia as similar to sharing a boat with a tiger. This figuration allowed me to theorize from a space outside of academia in articulating the precarious nature of my position: that at no point am I fully safe. I am always vulnerable to being prey, even when the predator is sleeping. That there is no actual shared space of interest convergence between the predator and the prey, although there might be a somewhat idealized desire for symbiotic survival. I became aware that the only interest served by my presence in the boat with the tiger is my survival. Everything I had done to keep the tiger fed, keep us afloat, and keep us away from disaster, was for the purpose of my survival; any shortcoming in my effort would have resulted in my being harmed.

Life of Pi, the movie, is also a deeply spiritual visual essay, the discussion of which is beyond the scope of this chapter. However, as someone who maintains a contemplative practice through meditation, mindful writing, and art-making, and engages in shadow work, I realized that in some form, trying to keep the tiger fed and satisfied also saved me from spiraling into deep depression and surrendering my own life energies. Dealing with various intersected structures of oppression in large and small ways is spirit and life-force draining. Therefore, although the tiger could have survived without me in the boat by finding someone else to prey on, in some way feeding the predator offered me an illusory escape from having to engage in shadow work that had the potential for paralyzing me with heaviness negating my will to live. In that way, however oppressive it has been to feed the tiger, it has also allowed me to incrementally pace my contemplative shadow work and be present for my survival, and by extension, life.

I titled the theorizing of my experiences *Acrobatic Writing: Feeding Fish to Richard Parker* and identified the theoretical and methodological moves I made in constructing the paper as well as navigating academia. For example, I acknowledged the heavy burden I carry of being a Third World Broker, despite my consistent protests, to constantly translate knowledge from Brown-embodied locations. Whether I locate my work or not, I am always already a Third World Broker, and therefore I am complicit in various in/visible ways in producing and proliferating colonizing discourses. I also noted that in a move to unlearn the colonizing effects on/of my work, I began to imagine that the primary audience for my work was my participants, transnational women dis/located in higher education in multiple parts of the world. This move began to change the way I wrote and how I held myself accountable.

Using the figuration from *Life of Pi*, I found a fertile intersection of spaces that were de/colonial, creative, and contemplative. Further, since the movie (based on a novel) lies outside the discourse that is privileged in my disciplinary spaces, it was important for me to use the figuration to create expansive and dialogic possibilities. Working through the theorizing, I was able to understand the tensions of being in a boat with a tiger, and how in some way such trapping has kept me motivated to tap into my life source energies instead of fully surrendering to the burden of daily and intergenerational oppression within and beyond academia. Seeking creative and contemplative spaces allowed me to journey deeply within, uncovering vulnerable, raw, and emotionally exposed parts of myself. In doing so I tried to legitimize the necessity of such moves in knowledge production and demonstrate why they must not be dismissed using colonizing and other disciplinary apparatuses.

Until recently, I believed that de/colonizing ontoepistemologies or approaches to inquiry must always exist in relation to colonization, informed by the framing of those who understand the liminality and hybridity of shuttling between multiple subject positions. However, as I continued to look for ways to expand my academic horizons, seeking knowledge-making sources beyond academia, I began to search for examples of de/colonizing thinking among those who fought the British when they ruled in India for 300 years. My need to find sources outside of academic gatekeeping led me to a children's writer, Sukumar Ray.

Ray's stories can be considered absurdist, nonsensical work. Their weak parallel in the west might be stories like Lewis Carroll's *Alice's Adventures in Wonderland*, though I would argue that Ray's work lacked even the modicum of coherence found in Carroll's book. In my recent writing, I described the context of one of Ray's works as follows:

> At the start of the 1919 harvest season (Baisakhi[1]) in British-occupied India, thousands of Sikh visitors made a pilgrimage to offer prayers in Jallianwala Bagh, a public garden in Amritsar with a walled exterior and five exits. The visiting villagers were unaware that General Dyer of the British Indian Army, attempting to prevent organized resistance, had banned all assemblies. Dyer ordered Indian soldiers under his command to shoot their own people, who were unarmed and unable to escape, as all exits were blocked. Thousands were murdered that day (Tharoor, 2017). Two years later, in 1921, Bangali[2] author Sukumar Ray published a children's book entitled *HaJaBaRaLa*.
>
> (Bhattacharya, In PressB, p. 5)

HaJaBaRaLa is a nonsensical word, mimicking the sounds of the Bangla alphabet similar to H, J, B, R, and L. The story is without plot, and features a ten-year-old boy who dreams about a handkerchief turning into a cat; a talking crow who does irrational math calculations; a part-alien, part-human, part-monkey, part-owl hybrid character, Hijibijbij, who shares nonsensical jokes and reflections and rolls around on the ground laughing; and a sleeping owl who serves as a judge in a court proceeding, among others (see Figure 8.3).

Bhattacharya, K. (In Progress). Exposing the fault lines of oppression: Par/Desi narratives of South Asian experiences in higher education. *International Journal of Qualitative Studies in Education.*

FIGURE 8.3   Absurdity and nonsense in play

Ray created not only a nonsensical story, but also nonsensical words for which there was no translation in any language. Consequently, the text was not intelligible to the British. To the Bengali people for whom the text was written, however, it became a treasure because it offered joy, whimsy, and play. Providing escape from the daily brutality of British oppression, the text did not once mention the British or colonization. Instead it created an utter absurdity in dreamspace. Ray's nonsensical words became idiomatically meaningful in the Bangla language over the years as parents read this text to their children over generations, eliciting joyful giggles.

At a time when the British killed for no reason, writing a subversive text that disrupted the order of things, including sense-making, and created chaos in dreamspace was an action that lay beyond disciplining. What was there for the British to discipline? Or more accurately, how were they to know that Ray's text would become a preserver of language and a force of spirit-building resistance over the years? Ray's work held no hint of resisting the British, nor of resisting oppression in any form.

Instead, Ray's characters interacted through made-up words and otherworldly concepts. Whereas colonization presents a clear and direct threat that erases language and heritage, Ray's work has stood the test of time, protecting language and heritage via intergenerational storytelling. Many years later, parents in India and across the world continue to read Ray's text to their children, while their children use the text to honor their parents after they pass away, and share it with the next generation.

This creative and subversive text was anti-colonial without marking itself as such, thereby circumventing a relationship to colonialistic thinking. Instead, by moving to a completely absurdist space, the text created ways of knowing and being that prevented the spirit-draining that occurs while living with/in colonial oppression. In a context in which such oppression is normalized and cannot be seen for the absurd state that it is, Ray's absurdity provided much-needed relief from which people could draw strength.

Using Ray's work, I revisited the data from my dissertation and created absurdist narratives by bringing in Ray's characters and placing them in dialogue with my study participants and myself. The characters serve as foils to the central protagonist, stand in as ethical benchmarkers, or reveal aspects of a situation that I might not have thought about (Bhattacharya, 2018a). In the past I had conceptualized de/colonizing as an entangled and interrelated concept of resistance and freedom. Without fully abandoning this conceptualization, I have drawn on Ray's work to expand my understanding of incorporating absurdity as a legitimate form of inquiry, analysis, and representation that is perhaps counterintuitive to research. This form of theorizing from the streets is an invitation to imagine, play, and be expansive, rather than accepting the limitations of the binary relations of resistance and freedom, the oppressor and the oppressed.

Benchmarks

As evidenced in multiple discourses in qualitative research, established benchmarks, standards, and other measures of rigor, goodness, or quality are often rife with contestations, paradigmatically and theoretically driven, and in many cases completely discredited (Cannella & Lincoln, 2004; Freeman, deMarrais, Preissle, Roulston, & St. Pierre, 2007; Tracy, 2010). Therefore, as I proceeded with a de/colonial, contemplative, and creative approach, I remained open to the idea that practice is inquiry, and in some cases a necessary path. Throughout this chapter, I have explained why and how my work may be considered theorizing from the streets. Where possible, I have described the creative, de/colonial, and contemplative moves I have made. In this section I highlight how I engage with certain points of consideration to identify benchmarks for my work.

First, when engaging with contemplative and de/colonizing work, I look for ways to be expansive and generative. This means I sometimes use art-making as practice before considering research purpose, methodology, or any other design agendas. I also look towards what would be spirit-nurturing in order to draw strength from it, especially if I am navigating multiple spirit-draining terrains. Colonialism is abuse, and being in relationship with colonialism is akin to being in a relationship with a narcissistic, abusive partner. Moreover, structures of oppression work in tandem to normalize oppression and create an absurd reality for those who are oppressed, as if they have few or no choices. Expansive thinking and spirit-nurturing processes in inquiry represent an unlearning of these discursive and material realities and a refusal to accept such realities as normal, logical, or rational.

Second, engaging in such refusal calls for creating our own forms of absurd practice that are intentionally nonsensical, imaginative, playful, and creative, thereby opening a portal for inner journeying or even journeying to other realms of being and knowing. Absurdity and liminality are freeing and give us permission to return to a state of being that rejects restrictive boundaries both within and beyond academia. Third, while one could argue that any practice of mindful awareness of self and self in relation to other and the world is contemplative, here I would add that engaging in self-excavation and shadow work are critical and necessary contemplative practices.

Fourth, as noted previously, the practice of research should integrate folk, communal, and indigenous wisdom instead of co-opting, appropriating, and obscuring them or rendering them backwards, inappropriate, or invisible. In this work I have demonstrated how I sought out these sources of wisdom to inform my journey of un/learning. If we are to become expansive, we must look beyond the academy to find inspiration for our theory, methodology, and practices of inquiry. We must draw on sources as varied as literature, music, poetry, drama, art, comedy, movies, conversations with people on the streets, and so on, to consider how our work can exceed the boundaries of academia. I remain uninspired to produce work that is accessible only to those who read and think the same things we do. Finally, if our communities and the participants in our inquiry are not the intended audience for our work, we are guilty of inflicting grave ontoepistemic, discursive, and material violence on them.

In Closing

By tracing my academic journey, in this chapter I demonstrated how I engaged in un/learning de/coloniality, specifically through creative and contemplative approaches in arts-based qualitative inquiry. I discussed the impossibility of imagining a pure decolonial space, yet also shared an example of what such a space could look like when intersected with creativity and contemplation, by highlighting a text from a children's author who wrote during the British Raj in India. Generally, I remain outside, on the margins of the privileged discourses in qualitative research. Yet, I cannot dismiss the spacemaking that the invitation to write this chapter creates, for which I am grateful and optimistic of our collective futures in qualitative inquiry. Even though I find it difficult to create entry points in privileged discourses, I would forego a belongingness in favor of a freedom-inducing space of imagination, play, and absurdity.

Notes

1 Inspired by Gloria Anzaldúa, I do not italicize non-English words, to disrupt anglo-centric ways of reading.
2 Bangali refers to a person from West Bengal, India, or Bangladesh. I use Bangali, instead of Bengali, because it is how the author would identify himself. *Note*: Bangla is the language spoken by those who identify as Bangali.

References

Anzaldúa, G. (1987/1999). *Borderlands la frontera: The new mestiza* (Second edn). San Francisco: Aunt Lute Books.

Anzaldúa, G. (2015). Let us be the healing of the wound: The Coyolxauhqui imperative–La sombra y el sueño. In A. Keating (Ed.), *Light in the dark: Luz en lo oscuro* (pp. 9–23). Durham, NC: Duke University Press.

Barad, K. (2007). *Meeting the universe halfway: Quantum physics and the entanglement of matter and meaning*. Durham, NC: Duke University Press.

Bhattacharya, K. (2005). *Border crossings and imagined nations: A case study of socio-cultural negotiations of two female Indian graduate students in the U.S.* (Ph.D. Dissertation), Athens, GA: University of Georgia.

Bhattacharya, K. (2007). Consenting to the consent form: What are the fixed and fluid understandings between the researcher and the researched? *Qualitative Inquiry, 13*(8), 1095–1115.

Bhattacharya, K. (2009). Othering research, researching the Other: De/colonizing approaches to qualitative inquiry. In J. Smart (Ed.), *Higher education: Handbook of theory and research* (Vol. XXIV, pp. 105–150). Dordect, The Netherlands: Springer.

Bhattacharya, K. (2016a). Dropping my anchor Here: A post-oppositional approach to social justice work in education. *Critical Questions in Education, 7*(3), 197–214.

Bhattacharya, K. (2016b). The vulnerable academic: Personal narratives and strategic de/colonizing of academic structures. *Qualitative Inquiry, 22*(5), 309–321. doi:10.1177/107 7800415615619.

Bhattacharya, K. (2018a). Exposing the fault lines of oppression: Par/Desi narratives of South Asian experiences in higher education. Paper presented at the American Educational Research Association, New York, NY.

Bhattacharya, K. (2018b). Walking through the dark forest: Embodied literacies for shadow work in higher education and beyond. *The Journal of Black Sexuality and Relationships, 4*(1), 105–124.

Bhattacharya, K. (In PressA). Contemplation, imagination, and post-oppositional approaches in qualitative inquiry. *International Review of Qualitative Research, 11*(3).

Bhattacharya, K. (In PressB). Nonsense, play, and liminality: Putting post-intentionality in dialogue with de/colonizing ontoepistemologies. *Qualitative Inquiry*.

Bhattacharya, K., & Payne, R. (2016). Mixing mediums, mixing selves: Arts-based contemplative approaches to border crossings. *International Journal of Qualitative Studies in Education, 29*(2), 1100–1117. doi:10.1080/09518398.2016.1201163.

Braidotti, R. (2013). *The Posthuman*. Cambridge: Polity Press.

Cannella, G. S., & Lincoln, Y. S. (2004). Dangerous discourses II: Comprehending and countering the redeployment of discourses (and resources) in the generation of liberatory inquiry. *Qualitative Inquiry, 10*(2), 165–174.

Freeman, M., deMarrais, K., Preissle, J., Roulston, K., & St. Pierre, E. A. (2007). Standards of evidence in qualitative research: An incitement to discourse. *Educational Researcher, 36*(1), 25–32.

Goffman, E. (1959). *The presentation of self in everyday life.* New York, NY: Anchor Books.

Jung, C. G. (2014). *Collected works of C.G. Jung, volume 9, part 1: Archetypes and the collective unconscious* (G. Adler & R. F. C. Hull, Trans. 2nd edn). Princeton, NJ: Princeton University Press.

Smith, L. T. (1999/2012). *Decolonizing methodologies: Research and indigenous peoples* (Second edn). London: Zed Books.

Spivak, G. Chakravorty (1993). Can the subaltern speak? In P. Williams & L. Chrisman (Eds.), *Colonial discourse and postcolonial theory* (pp. 66–111). New York, NY: Columbia University Press.

St. Pierre, E. A. (2013). Post qualitative research: The critique and the coming after. Paper presented at the Ninth International Congress of Qualitative Inquiry, Champaign/Urbana, IL.

Tharoor, S. (2017). *Inglorius empire: What the British did to India.* Brunswick, Victoria: Scribe Publications.

Tracy, S. (2010). Qualitative quality: Eight "big-tent" criteria for excellent qualitative research. *Qualitative Inquiry, 16*(10), 837–851.

9

STAY HUMAN

Can We Be Human after Posthumanism?[1]

Svend Brinkmann

Prologue

One day in the fall of 2017 I was sitting in my university office reading about "posthumanism" and "post-qualitative research." Like many other qualitative researchers, I had for a long time been critical of an essentialist humanism that seemed to privilege a specific kind of subject (typically male, white, middle class and from the imagined hemisphere we call "the West"), conceiving of other kinds of subjectivity as deviations from this essential one, and I had thought a lot about how to develop new practices of inquiry—perhaps to be called post-qualitative ones—that did not automatically proceed from this default position. It was inspiring to get acquainted with post-qualitative philosophy and research and the whole posthuman critique of key qualitative terms such as experience, voice, subjectivity, and investigative and analytical practices such as interviewing and coding (St. Pierre, 2011). However, when I left my office to get a cup of coffee on that day, I saw a poster on the wall inviting me to an open dialogue about refugees; a discussion that had really taken off in my country, Denmark, in 2015 with the so-called "refugee crisis" in Europe that had witnessed large numbers of people crossing the Mediterranean from North Africa and the Middle East, fleeing from war-stricken Syria and elsewhere. The headline of the poster said: "Stay human!" Amnesty International and other NGOs were listed as supportive of the dialogue about the politics of integration and multiculturalism.

I felt a kind of cognitive dissonance. The contrast between the call to stay human and the arguments in favor of posthumanism was striking. Suddenly, the philosophical and methodological reflections from my office readings seemed rather disconnected from the human tragedies among refugees and immigrants. It suddenly felt like an intellectual luxury to play around with notions of humanism and posthumanism, and I walked back to my office with a sentiment of disillusion.

In such situations of dissonance, one may either revise one's beliefs—in this case about the importance of philosophies of (post)humanism—or one may change one's courses of action, e.g., by showing up at the meeting and engaging in voluntary work or something like that. I decided to do something in between, namely, continue studying the problematics of humanism and the critique thereof, but at the same time trying to juxtapose this discussion with the ethical and political problems of the world today, where an idiom of humanism still serves a purpose that most of us find laudable (I do, at least), summarized in the simple imperative of the poster: Stay human. What does it mean to stay human? Should we stay human if "the human" (whatever that is) is partly responsible for our problems? And if we are to stay human, then how can we do that without sliding into problematic essentialisms? In other words, who can define what "the human" is in a world that is moving toward posthumanism?

I cannot promise to answer all these difficult questions here in this short text. I mention them only to indicate the scope of the discussion, and I hope to be able to address them in various ways in the years to come. Here, I will proceed by first briefly characterizing posthuman philosophy and the related post-qualitative critique of conventional qualitative research (based primarily on Brinkmann, 2017), before, in a second phase, discussing if there is something worth preserving in "humanism" (and whether this is even intellectually feasible now), and, third, finally borrowing from the work of philosopher Hans Jonas, who was trying to articulate a kind of humanism that avoids essentialism and the dangers of anthropocentrism. Perhaps something like that is what we need to stay human without becoming dangerously anthropocentric.

Post-Qualitative Research and Posthumanism

Building on posthuman and new materialist philosophies of thinkers such as Karen Barad (2007), Donna Haraway (1991) and Gilles Deleuze (Deleuze & Guattari, 1987) among many others (including, not least, the whole oeuvre of Michel Foucault), there is now an increasingly prevalent post-qualitative critique of what the "post" scholars summarize as "conventional qualitative research" (see e.g., Aghasaleh & St. Pierre, 2014). The critique amounts to a political, ethical, ontological and methodological questioning of the assumptions and practices of humanistic qualitative inquiry as normally practiced (e.g., in the form of interviewing and participant observation and paradigms such as phenomenology, hermeneutics and ethnomethodology). The main argument of the post-qualitative scholars is that what we have come to know as qualitative research springs from a modernist humanism that ought to be abandoned, because it is both ethically and ontologically problematic. As Patti Lather and Elizabeth St. Pierre, two of the most central persons in the post-qualitative movement, put it: "rethinking humanist ontology is key in what comes after humanist qualitative methodology" (Lather & St. Pierre, 2013, p. 629). Onto-logy is about "what there is," and this, according to the post-qualitative researchers, cannot be comprehended by using mechanical models based on Newtonian physics and the scientificity constructed on its basis. Often, the quantum theory of Niels Bohr is invoked as providing a different ontological ground than Newton's mechanical physics (especially by Barad, 2007). "What there is" also cannot be comprehended, according to the post-qualitative per-spective, by focusing on the subjective or experiential side of the divide that came into being historically with the dichotomy of the objective and subjective that resulted from modernist science. In this light, qualitative research is simply a faulty enterprise, the argument goes, because it concerns itself with just one side of a false dichotomy. It privileges the field of human experience and sub-jectivity, but our ways of thinking of this field – and the techniques and methods developed to study it – presupposes a metaphysics that was misguided from the very beginning. As Lather and St. Pierre put it:

> If we cease to privilege knowing over being; if we refuse positivist and phenomenological assumptions about the nature of lived experience and the world; if we give up representational and binary logics; if we see lan-guage, the human, and the material not as separate entities mixed together but as completely imbricated "on the surface"—if we do all that and the "more" it will open up—will qualitative inquiry as we know it be pos-sible? Perhaps not.
>
> *(Lather & St. Pierre, 2013, pp. 629–630)*

According to these post-qualitative scholars, refusing the dichotomy between the objective (quantitative positivism) and subjective (qualitative phenomeno-logy of lived experience) means that conventional qualitative inquiry is no longer possible.

Referring to the helpful programmatic text of St. Pierre, Jackson and Mazzei (2016), we can summarize the main philosophical ideas behind post-qualitative thinking along three lines: First, matter (or nature) is understood as agentic and always changing. In a way, this is an ancient idea in philosophy, going back to pre-Socratic philosophers of flux such as Heraclitus, who depicted the world as a constant flow of becoming. Matter is not simply cold and dead, to be studied by mechanical sciences, but a warm and vibrant process that acts and develops. Along the same lines, Actor-Network theorists such as John Law and Bruno Latour have granted agency to non-human "actants" within networks of prac-tices (Latour, 2005), and in his classic *We Have Never Been Modern*, Latour argued that the separation of nature (of which we may allegedly obtain objective knowledge) and society/culture/the human (for which we need qualitative approaches of phenomenology and hermeneutics) was never possible, although this was the backbone to the Enlightenment and the whole modern project (Latour, 1993), *and* that from which the notion of qualitative research historically emerged. All in all, this first point deconstructs the opposition between a sphere of passive, inert matter on the one hand and a sphere of meaningful human experiences, discourses and actions on the other. Some follow Haraway (1991) and talk about an "entanglement" of the material and the semiotic— human living, thinking and acting is always also material, just as the material is always also semiotic.

Second, as argued by St. Pierre and colleagues, there is "a heightened curiosity and accompanying experimentation" concerning existence among post-qualitative researchers (St. Pierre, Jackson & Mazzei, 2016, p. 102). This means that thinking in the post-qualitative movement goes on not just about "what there is" but more so about what may become (or, to put it in more post-qualitative terms: "what there is" *is* just a process of becoming). As Foucault was fond of pointing out, the goal should not be to discover who we really are, or realize our true, humanist, authentic selves—for these are illusions—but rather to refuse who we are and suggest alternative forms of existence (Foucault, 1988). Post-qualitative research continues to some extent the playful experimentation that we have come to associate with postmodernism and practices such as investigative poetry, arts-based research and creative analytical practices in general (Richardson & St. Pierre, 2005), which are also practiced within more theoretically mainstream approaches, but with more emphasis on playing with theory and philosophy. Post-qualitative work begins not with method, but with (posthuman) theory. Theorizing is seen as generative, and there is a veritable semantic explosion of new words and concepts with the stated ambition of eroding the established binaries that are seen to be inherent in qualitative research.

Third, there is a general critique and rejection of the philosophy of representation. The "posts," argues St. Pierre, "announce a radical break with the humanist, modernist, imperialist, representationalist, objectivist, rationalist, epistemological, ontological, and methodological assumptions of Western Enlightenment thought and practice" (St. Pierre, 2011, p. 615). What Rorty attacked as "the mirror of nature" some decades ago (Rorty, 1980), i.e., the (mistaken) idea that the human mind is a representational device that may mirror—if the proper methods are used—a world that is independent of the mind, is completely dismantled, for there is no detached human being (or metaphorical mirror). This might seem to be a form of social constructionism, but that is a mistaken interpretation, according to the posthumanists, for although social constructionists are aligned with posthumanism concerning their shared critique of essentialism and experientialism, social constructionism lives off the same modernist separation of matter and meaning as conventional qualitative research and simply focuses on the latter (or, in radical versions, claim that there *is* nothing else—given their belief that the material is also a social construction).

Latour for one has been very clear that he dislikes the social constructionist co-option of Actor-Network-Theory, given that they (Latour and the constructionists) operate with two very different ontologies.

Furthermore, the deconstruction of representationalism is also thought to lead to new kinds of ethics in the post-qualitative movement: St. Pierre, Jackson and Mazzei talk about how post-qualitative research implies an "ethico-onto-epistemology, which makes it clear that how we conceive the relation of knowledge and being is a profoundly ethical issue, as is the relation between the human and the nonhuman" (St. Pierre, Jackson & Mazzei, 2016, p. 99). And further:

> If humans have no separate existence, if we are completely entangled with the world, if we are no longer masters of the universe, then we are completely responsible to and for the world and all our relations of becoming with it.
>
> (p. 101)

However, there is little in the post-qualitative critique to help us navigate these ethical issues of responsibility, and it also seems logically possible to draw the exact opposite conclusion: If we are no longer masters of the universe, but simply a Deleuzian "fold" or circulating affects, then we have absolutely no ethical responsibility, because there is ultimately nothing we can control. This discussion is likely to be significant among post-qualitative scholars in the years to come.

Following from the processual and agentic view of reality, the heightened curiosity and experimentation, and also the critique of representationalism—in short the dismantling of what is otherwise called humanism—comes logically a rejection of "humanist methodology": "The failure of the humanist subject produces the failure of humanist methodology" (St. Pierre, 2011, p. 618). St. Pierre provides a list of conventional qualitative terms that have been deconstructed by post-qualitative scholars, including *interview, validity, data, voice* and *reflexivity* (p. 613). These concepts are said to spring from the notion of the bounded human subjective *self* that should conventionally be called forth through *interviewing*, for example, giving *voice* to the individual and resulting in *data* that should be *coded* and *analyzed* by a separate researcher who needs to engage in a *reflexive process* in order to be clear about her *subjective standpoint*. Something like this, in short, is what post-qualitative researchers frame as conventional qualitative research in a humanist vein, and what they find deeply problematic.

Toward a Humanism after Posthumanism

I hope the reader senses that I have much respect for the philosophical project behind post-qualitative research. I have tried to characterize it above in a way that does justice to the richness and sophistication of its ideas. But I wish to raise the question whether there could be reasons for holding on to a form of humanism and corresponding qualitative research practices now that we are after the post-qualitative critique? I will propose here a number of basic reasons to hold on to what Marecek (2003) has called "a qualitative stance" that rests on humanist assumptions. Marecek talks about a qualitative stance for psychology specifically, but I believe the argument has a wider scope. In short, I think we can (and should) take a humanist stance that at once recognizes the ontology of flux—and the whole entanglement of the material and the semiotic—but which, exactly because of this recognition, posits the need for building more stable practices (including investigative practices) around the human. This might be how to stay human in a posthuman world.

As I touched upon at the beginning, the world is now witnessing a number of crises that transcend national and geographical borders. The economic crisis, connected to a defective neo-liberal system that leads to poverty and increased inequality; the climate crises, giving rise to new conflicts and struggles over scarce resources; and the refugee crisis, spawned by wars and vexed political interests. It is clear, as Latour (1993) pointed out years ago, that such crises cannot be understood exclusively either in terms of "the objective" (with reference to the workings of the natural world alone) or "the subjective" (with reference to the workings of the human and social world alone). However, if we return to the latter example of the refugee crisis, it still seems to me that we need the concept of humanism, not least for rhetorical reasons, if we want to argue that ethical and humanitarian—and not only economic—arguments should play a role in how to address the problem. In my own country, Denmark, which for decades has prided itself on being something of a "humanitarian superpower" (in spite of its small geographical size), the concept of humanism has very rapidly been discursively transformed in the public debate from signaling a worthy ideal to being something like a term of abuse. In public discussions, a humanist is now seen as someone who is irresponsible and out of touch with reality, because—as it is often said—"we cannot receive all the refugees in the world." We now have a frightening climate of political posthumanism or even anti-humanism, and although this is quite a different argumentative context than the philosophical ideas of posthumanism, I believe that it is a sign that the idea of humanism is ethically too important for us to lose. It does point to something valuable when people urge us to "stay human!"

As we saw above, humanism has been attacked by the "post" thinkers exactly for ethical reasons: "recent attacks have denounced 'humanism' as a form of white, male, Western, elite domination and colonization that is being imposed throughout the world and that brings with it too strong a sense of the unique individual," Plummer observes (2011, pp. 199–200). In spite of this, however, Plummer defends what he calls "a more complex humanism" that sees human beings as "always stuffed full of their culture and history" and as "both embodied, feeling animals and creatures with great symbolic potential" (p. 200). From my perspective, this seems promising: To see human beings as ontologic- ally "stuffed full" of culture and history (meaning) *and also* corporeality and ani- mality (matter), and yet—with and within all this stuff—arguing that humanism is worth advocating as an ideal.

Some critics would here accuse humanists of "speciesism," in other words, granting rights and dignity to humans on the basis of species membership alone. However, that would amount to accusing ants of "Formicidaeism" (the ant equivalent to humanism). Of course, ants cannot help but see the world from the ant perspective. They have a special interest in their fellow ants and the flourishing of their anthills, just as humans have a special interest in their fellow humans and the flourishing of their communities. And if anti-human critics were to object that this comparison cannot be made, because there is a differ- ence between ants and humans related to the wider human capacity for moral reflection and responsibility, then they would paradoxically have confirmed the humanism that they otherwise criticize! Yes, this difference between humans and other animals certainly seems to exist, one could say, and the fact that humans can be humanists because of their special skills and capabilities does not mean that they are not animals that live in a natural-social world that is con- stantly changing. But it does mean that this is simply part of the human condition.

Although such a reply could agree ontologically with the flux thinkers, it stands opposed to their normative valuation of the unstable, as articulated, for example, by St. Pierre, who writes: "My desire is for post inquiry to remain *unstable* as we create different articulations, assemblages, becomings, mash-ups of inquiry given the entanglement that emerges in our different projects" (St. Pierre, 2011, p. 623). But is this really what is needed today? Marx and Engels famously wrote in *The Communist Manifesto* that "all that is solid melts into air" because of the

> constant revolutionizing of production, uninterrupted disturbance of all social conditions, everlasting uncertainty and agitation [that] distinguish the bourgeois epoch from all earlier ones. All fixed, fast-frozen relations, with their train of ancient and venerable prejudices and opinions, are swept away, all new-formed ones become antiquated before they can ossify.
>
> *(Marx & Engels, 1848)*

The destabilizing tendencies of the capitalist system have only increased since the middle of the nineteenth century, most recently with the epoch variously named late capitalism, post modernity or liquid modernity (Bauman, 2000), which advances a kind of network capitalism (Brinkmann, 2008). The notion of the network (cf. Latour) or the rhizome (cf. Deleuze) figure prominently in the new materialist ontologies. Various scholars, however, are now problematizing the ideological effects of this discursive formation. From the perspective of critical theory, for example, Hartmann and Honneth have argued that "to the extent that the image of a society pervaded by networks takes hold as a fundamental means of societal self-description, other images of the social whole lose in influence" (Hartmann & Honneth, 2006, p. 52). French pragmatists Boltanski and Chiapello (2005) have concluded that the notion of network both as an organizational and institutional mechanism, and also as an ideology, has become central in postmodernity, characterized by neo-liberal flexibilization. They argue that the institutionalization and ontologization of networks has few chances of leading to social justice, because networks do not consider those who find themselves 'disconnected' or on its margins. One is either part of the networks and must function according to their (capitalist) logics, or one is outside the networks and consequently socially impotent. In their analysis of "network capitalism," Hartmann and Honneth agree with this diagnosis: "In network capitalism, on this thesis, citizens tend to perceive their efforts, successes, and failures as individualized, so that a reference to the greater whole scarcely seems possible any longer" (Hartmann & Honneth, 2006, p. 52). In his book on Deleuze, Slavoj Žižek has gone so far as to directly criticize Deleuze—this key posthuman thinker and one of the most important sources of inspiration for the post-qualitative scholars—for being "the ideologist of late capitalism." It is worth quoting Žižek at some length in order to understand this harsh verdict:

> There are, effectively, features that justify calling Deleuze the ideologist of late capitalism. Is the much celebrated Spinozan *imitation afecti*, the impersonal circulation of affects bypassing persons, not the very logic of publicity, of video clips, and so forth in which what matters is not the message about the product but the intensity of the transmitted affects and perceptions? ... Is this logic in which we are no longer dealing with persons interacting but just with the multiplicity of intensities, of places of enjoyment, plus bodies as a collective/impersonal desiring machine not eminently Deleuzian?
>
> *(Žižek, 2012, pp. 163–164)*

This critique of Deleuze in effect claims that his rhizomatic network ontology mirrors and facilitates the ideology of late capitalism. As I see it, advocating humanism, the institutions that grant rights and responsibilities to human beings, *and* the whole qualitative interest in human lives and experiences is needed to counteract this ideology. Transforming an ontological realization of flux into an advocacy for instability and the impersonal circulation of affect may simply come to serve some of the most problematic tendencies of our times—and contribute to melting the few remaining solids into air.

Humanism without Anthropocentrism?

Can "a more complex form of humanism" that Plummer talked about be invoked to counter the destabilizing tendencies of late capitalism? And if so, where can we turn to find such humanism that does not become anthropocentric and destructive? I have recently become interested in Hans Jonas' philosophy because it on the one hand represents a version of the "new materialisms" (represented today by names such as Barad, Haraway and Latour), but on the other hand maintains a distinct human (and humanist) perspective without thereby reducing itself to anthropocentrism. Like the new materialisms, Jonas' philosophy is starkly anti-dualistic—it dissolves all absolute distinctions between nature and culture, mind and matter—and presents a philosophy of continuity between plants, animals, and human beings. This is much needed in an age of the Anthropocene (the geological term for the current time when human activity transforms the planet) and in a world of crises (global warming, reduced biodiversity, etc.) (Davies, 2016).

Jonas was a pupil of Martin Heidegger, and, as a Jew, he had to flee from Nazi Germany before World War II, first to England and Palestine, and then, as a soldier in the British army, to Italy and finally to Germany again as part of the victorious army (he had sworn to return to Germany only as a victor). After learning that his mother had been killed in Auschwitz, he decided never to live in Germany again, and he eventually became a professor at the New School for Social Research in New York City, where he lived for the rest of his life (Jonas died in 1993 at the age of 89).

Although Jonas' work is known in certain scholarly circles, he is not widely read today. This is a shame, since he was one of the first scholars to address the dangers of technological development and climate change in great depth. But in the heydays of postmodernism and discourse analysis, when much was destabilized through deconstruction, he was probably too metaphysical and "materialist," and now that the pendulum has swung towards new materialisms, he is perhaps too easily (and mistakenly) seen as anthropocentric.

Like other students of Heidegger and post-Heideggerian thinkers (Arendt, Gadamer, Levinas), Jonas developed a moral philosophy based on the experiences of the Holocaust, partly as a reaction to what he saw as the lurking nihilism in Heidegger's existential phenomenology (Jonas, 1996). I find that Jonas' work is particularly compelling in the era of the Anthropocene. In a recent book on this geological epoch, where humans are a decisive force in transforming the planet, Davies articulates the normative demand issuing from the realization of what he calls deep historical time: The new epoch offers a way for environmentally conscious citizens "to see themselves as 'members of deep time, along with trilobites and Ediacaran organisms,' as 'one expression of the ever-evolving planet'" (McKay quoted in Davies, 2016, p. 193). Without referring to Jonas, this is actually a quite precise characterization of his work and his attempt to provide "an existential interpretation of biological facts" (which was how he often referred to his own work).

Centrally, for Jonas (1984), axiology is a part of ontology (p. 79), which is to say that values emerge from nature itself. He believed that ethics should be based not on God, community or the self, but on the nature of things. This is already to break with anthropocentric ethics that find the source of value in the human realm of feelings (e.g. utilitarianism) or the will (e.g. Kantianism). In a way, it is to return to a Greek conception of an ontic logos as the ground of ethics (see Taylor, 1989), but in a way that is deeply informed by Darwin's theory of evolution and modern science. Unlike (some) posthumanist thinkers that find inspiration in quantum physics, Jonas turns to biology to understand a universe in flux. His philosophy of life, of biological facts, locates value in the very nature of life itself, even down to metabolism.

But what kind of value, what kind of normativity, then emerges from life itself? Jonas' answer was that all organisms are ends in themselves, because all have needs (Jonas, 1966, 1996). There is purposiveness in nature, he argues, which implies that being is normatively better than non-being. His philosophy of nature begins not with dead matter (leading to the riddle of Newtonian natural science about how life can emerge from death), but rather with living organisms in their environments and worlds. But, Jonas argues, humans are unique because they represent the maximal actualization of the purposiveness of nature (Jonas, 1996, p. 16). This is so because only humans are responsible for their lives and what they do. Jonas develops a philosophy of natural responsibility by first interpreting nature existentially (Jonas, 1966) and then developing an account of human nature specifically that centers on a metaphysical grounding of our ethical obligations (Jonas, 1984, 1996). Nature is not just a resource that is ready to answer to human needs and interests; it harbors intrinsic value. But humans have evolved to be creatures that can be responsible—like no other known organism—giving rise to what he called "the imperative of responsibility," which is really the imperative to secure the continued existence of life, human and otherwise, on a planet in jeopardy.

This may seem like anthropocentrism—seeing everything in the universe from a distinct human perspective—but in fact it is the opposite, and Jonas was insistent in his critique of anthropocentric ethical systems. What for Jonas was a fact—that there is "an unconditional duty for mankind to exist" (Jonas, 1984, p. 37)—arises not from human dominance over nature, but rather from life itself, since nature has produced a being that is actually able to assume responsibility, not just for itself, but for life as such.

In this context, I cannot of course explicate fully Jonas' complex theory of responsibility, or his advanced analysis of the continuity of being among plants, animals and humans (building on Aristotle, Hegel and Darwin), for the purpose is simply to emphasize how a deep thinker sought to demonstrate that ethics should rightly be seen as belonging to natural philosophy. The idea of anthropocentric ethics that a normative "ought" can only issue from humans themselves (from the will as in existentialism and Kantianism, or the emotions as in emotivism and utilitarianism) is not the result of a descriptive analysis, according to Jonas, but a metaphysical prejudice that can and should be discussed and criticized. Jonas believed that the deepest normative "ought" comes from life itself, from nature (and the nature of things). But this is not to say that humans have no role in ethical life. On the contrary, it means that humans have a special role to play as creatures endowed by natural processes with responsibility. Thus, Jonas formulates his "imperative of responsibility" as follows: "Act so that the effects of your action are compatible with the permanence of genuine human life," or, expressed negatively: "Act so that the effects of your action are not destructive of the future possibility of such life" (Jonas, 1984, p. 11). Needless to say, Jonas thought that we fail to pay heed to this imperative in the Anthropocene epoch.

All in all, Jonas develops a philosophy of life that gives humans a special duty to exist and protect the intrinsic value of nature. I have not unpacked all his arguments or his philosophical system here, but what I find interesting is how he articulates a philosophy that on the one hand builds on a relational, processual, anti-dualistic ontology like the posthuman and new materialist philosophers, but nonetheless—on the other hand—finds a way for us to stay human. Well, not just finds a way, but actually installs an imperative in us to stay human. We can and ought to be human without being anthropocentric. This is Jonas' message. And this has consequences not only for ethics, politics and our understanding of technology and nature, but also, I believe, for our investigative practices. Although Jonas was not in any way a (qualitative) methodologist, I believe there is in his work an implicit defense of an interpretative stance, a qualitative stance, to the understanding of life processes. After all, his credo calls for an existential interpretation of biological facts, and he meant "biological" in the broadest possible sense that include every human life process and experience. My guess is that he would find little reason to become post-qualitative, but rather ample reason to defend a humanist standpoint of ethical responsibility when considering the phenomena and problems of living.

Conclusions

The acting and experiencing human being is not an ahistorical, disembodied and universal intellect, but rather a historical, embodied, affective creature that lives in a sociomaterial world of flux and uncertainty. And yet, within that world, humans have managed historically to build relatively stable structures and routines together that make possible their experiencing and acting. Certainly, many of these structures and routines have been oppressive for various groups of people and animals, but simply advocating for destabilizing structures and routines as such—no matter what we talk about—seems to be a risky strategy for ethical and political reasons. For such structures are likely a precondition for the formation of human beings that are capable of taking responsibility and acting in ways that can be defended and justified ethically. In this way, we should think of stabilizing practices—including our scientific techniques and arts—as providing the precondition for ethical human life. Some of the stabilizing practices have been qualitative research practices, which study human lives and ways of experiencing and acting in their many forms. Taking what Marecek (2003) calls a "qualitative stance" toward human life means studying phenomena in a context of history, society and culture; it means resituating humans in their life worlds; and it means approaching humans as reflexive and intentional agents who can create meaning. I will argue that this represents a form of humanism that is worth holding on to, because it respects the idea of responsibility—including Jonas' imperative of responsibility. We should in my view appreciate the ontological theorizing conducted under the post-qualitative and post-humanist banners but see the precarious and unstable nature of reality as giving rise to an ethical demand for humans, viz. to enact relatively stable practices in which it becomes possible to conduct flourishing lives together and take the imperative of responsibility seriously. Qualitative research remains important as a set of investigative practices in this regard in our liquid modern world (Bauman, 2000).

Obviously, Jonas' philosophy does not provide us with an easy way to solve the refugee crisis or any of the other crises affecting the planet, but I believe it gives us some conceptual resources that allow us to remain humanists and use this vocabulary against xenophobic politicians, for example, in a way that at the same time preserves the insights of the new materialist thinkers. However, and needless to say, perhaps, the last word has not been said in this discussion, but my own best account is along the lines of Jonas—to strive for humanism without anthropocentrism. To stay human without being anthropocentric.

Note

1 This text is based on a presentation given at the International Congress of Qualitative Inquiry in Urbana-Champaign, Illinois, May 2018. That presentation was to some extent based on a previously published article (Brinkmann, 2017). The current text thus reworks materials from the presentation and the article, but set within a new framework, as part of an ongoing discussion of posthumanism and post-qualitative research.

References

Aghasaleh, R. & St. Pierre, E. A. (2014). *A reader's guide to post-qualitative inquiry proposals*. Retrieved from: http://goo.gl/3OC5b2.

Barad, K. (2007). *Meeting the universe halfway: Quantum physics and the entanglement of matter and meaning*. Durham, NC: Duke University Press.

Bauman, Z. (2000). *Liquid modernity*. Cambridge: Polity.

Boltanski, L. & Chiapello, E. (2005). *The new spirit of capitalism*. London: Verso.

Brinkmann, S. (2008). Changing psychologies in the transition from industrial society to consumer society. *History of the Human Sciences, 21(2):* 85–110.

Brinkmann, S. (2017). Humanism after posthumanism: Or qualitative psychology after the "posts." *Qualitative Research in Psychology, 14(2):* 109–130.

Davies, J. (2016). *The birth of the Anthropocene*. Oakland, CA: University of California Press.

Deleuze, G. & Guattari, F. (1987). *A thousand plateaus: Capitalism and schizophrenia*. Minneapolis: University of Minnesota Press.

Foucault, M. (1988). Technologies of the self. In *Technologies of the self*. London: Tavistock.

Haraway, D. (1991). *Simians, cyborgs and women: The reinvention of nature*. New York: Routledge.

Hartmann, M. & Honneth, A. (2006). Paradoxes of capitalism. *Constellations, 13,* 41–58.

Jonas, H. (1966/2001). *The phenomenon of life: Toward a philosophical biology*. Evanston, IL: Northwestern University Press.

Jonas, H. (1984). *The imperative of responsibility: In search of an ethics for the technological age*. Chicago, IL: University of Chicago Press.

Jonas, H. (1996). *Mortality and morality: A search for the good after Auschwitz*. Evanston, IL: Northwestern University Press.

Lather, P. & St. Pierre, E. A. (2013). Post-qualitative research. *International Journal of Qualitative Studies in Education, 26,* 629–633.

Latour, B. (1993). *We have never been modern*. Cambridge, MA: Harvard University Press.

Latour, B. (2005). *Reassembling the social*. Oxford: Oxford University Press.

Marecek, J. (2003). Dancing through minefields: Toward a qualitative stance in psychology. In P. M. Camic, J. E. Rhodes, & L. Yardley (Eds.), *Qualitative research in psychology: Expanding perspectives in methodology and design* (pp. 49–70). Washington, DC: American Psychological Association Press.

Marx, K. & Engels, F. (1848). *Manifesto of the Communist Party*. Retrieved from: www.marxists.org/archive/marx/works/1848/communist-manifesto/ch01.htm.

Plummer, K. (2011). Critical humanism and queer theory. In N. K. Denzin & Y. S. Lincoln (Eds.), *The SAGE handbook of qualitative research* (pp. 195–211). (4th edn). Thousand Oaks, CA: Sage.

Richardson, L. & St. Pierre, E. A. (2005). Writing: A method of inquiry. In N. K. Denzin & Y. S. Lincoln (Eds.), *Handbook of qualitative research* (pp. 959–978). (3rd edn). Thousand Oaks, CA: Sage.

Rorty, R. (1980). *Philosophy and the mirror of nature*. Princeton, NJ: Princeton University Press.

St. Pierre, E. A. (2011). Post qualitative inquiry. In N. K. Denzin & Y. S. Lincoln (Eds.), *The SAGE handbook of qualitative research* (pp. 611–625). (4th edn). Thousand Oaks, CA: Sage.

St. Pierre, E. A., Jackson, A. Y., & Mazzei, L. (2016). New empiricisms and new materialisms: Conditions for new inquiry. *Cultural Studies – Critical Methodologies, 16,* 99–110.

Taylor, C. (1989). *Sources of the self*. Cambridge: Cambridge University Press.

Žižek, S. (2012). *Organs without bodies*. London: Routledge.

PART III
Political Interventions

10

RESISTING THE COMMODIFIED RESEARCHER SELF

Interrogating the Data Doubles We Create for Ourselves when Buying and Selling Our Research Products in the Research Marketplace

Julianne Cheek

What This Chapter Is About

At the thirteenth International Congress of Qualitative Inquiry (ICQI) held at the University of Illinois in May 2017, I had the privilege of organising and being part of a session called 'Slowness, laziness, and stupidity: Antidotes to seemingly 'effective' scholarship and the neoliberal academy'. The aim of the session was to disturb what have become 'ordinary' neoliberally derived strategies and practices that qualitative inquirers are urged to embrace and enact daily. These include researching faster and working harder in order to produce more research 'output', calculated by the number of published papers and the amount of research funding won; and teaching smarter and more efficiently to produce more student completions in less time, with less workload.

There were four papers in the ICQI session, each of which deliberately used and main-framed a concept that would be taken as oppositional to certain core neoliberal premises: slowness vs. working faster (Koro-Ljungberg & Wells, 2017; Ulmer 2017); laziness vs. working harder (Gildersleeve, 2017); and being selectively stupid vs. working smarter (Cheek, 2017c). In this way the writers collectively hoped to challenge the reified status of practices commonly given mainframe status in neoliberal contexts; practices such as the ordinariness, even the possibility, of always working harder, faster and smarter. The aim was to expose and therefore make visible the extra-ordinariness of what has become normalised for many researchers as they struggle to produce more research in less time, supposedly working smarter by, for example, using technology to create more time to do more, while at the same time jostling for relative market position in the research marketplace.

Reflecting on the session, I was struck by three things that ended up providing the impetus and point of departure for this chapter. The first was that the neoliberal market-derived research marketplace in which, as researchers, we all find ourselves, was the elephant in the room in all the presentations. Even when not overtly named, it provided the context for understanding how the normalisation of assumptions such as the ordinariness and possibility of always working harder, faster and smarter has occurred, as well as what sustains such normalisation.

The second was that much of the discussion in the session could be viewed as a form of activism by individual researchers pushing back against such normalising tendencies. The deliberate use of terms such as *slow*, *lazy* and *stupid*, the antithesis of the *fast*, *hard* and *smart* mantras of neoliberal rhetoric, was part of this activism. Deliberately foregrounding such 'not normal' and devalued terms, and thereby not accepting the assumed and normal mainframe afforded to terms such as working *fast*, *hard* and *smart*, allowed different questions to emerge. For example, why should creating more research products in less time, using less effort, necessarily be equated with better and higher research performance? This, in itself, has been a fundamental premise when metrics are used to rank researchers against each other.

The third and, possibly, the most confronting thing that struck me as a result of reflecting on this session was that I might well be complicit, at least at times, in meeting (or at least cooperating with) the demands of this research marketplace and the commodification of my research self and my research products. This led to me having to reflect on questions that I often (in fact, nearly always) prefer to avoid. For instance, what was my motivation in applying for funding for research? Why did I choose to publish my work where I did, in the way that I did? Why do I know, track, and even want/need to know, what my research-related metrics are?

Thinking more about these three things and the connections between them, I realised that addressing whether or not I was complicit in taking for granted (or even actively participating in) the commodification of both myself and my research in the research marketplace, involved a process of questioning my own research-related actions *and* challenging those actions. I asked myself: what sort of researcher self am I? How did I come to be that self? Why did I choose to be that self? And, crucially, what self do I want to be, and what might this mean for the way that I think and undertake my research in the research marketplace?

What follows is a discussion of where this thinking has led me, and the possibilities it has opened up for me about how to embrace a form of activism, at the level of the individual researcher, to expose and thereby open up the possibility to resist the normalising tendencies of this research marketplace. The focus of the discussion is thus at the level of the actions that we as individual researchers take when we navigate and position ourselves, and our research, in the contested terrain of research metrics and research marketplaces. How do we do our research and present ourselves as researchers, and why do we do it this way?

I will begin by metaphorically inviting the 'elephant in the room', that is, the research marketplace, to introduce itself, to join us overtly, and thus to be a visible and explicit part of this discussion.

However, before I issue this invitation, I want to warn readers that this discussion can, and will, be a very uncomfortable one at times. This is precisely why it is not a discussion that we often have. It is much easier to talk about problems with metrics and research markets and how others use them, than to explore how we ourselves take a position in relation to, and make choices about, those metrics and markets. For as Fochler and De Rijcke (2017) note:

> The rise of new modes evaluating academic work has substantially changed institutions and cultures of knowledge production.... Many of us are experts on aspects of these changes. But at the same time, we too are part of the processes we are analyzing, and often criticizing.... This creates tensions that many of us reflect on.... Yet it seems that so far there has been little room in our field to reflect on and exchange this particular kind of experience-based knowledge. There are many different ways to engage with the dynamics of evaluation, measurement and competition in contemporary academia, or to play what we refer to colloquially here as the 'indicator game'.
>
> *(p. 21)*

The Research Marketplace – We Are All Interacting with It and Making Decisions about the Nature of that Interaction

Like it or not, as researchers we are all part of this neoliberally derived and metric-driven research marketplace. It is where researchers and others, such as funding bodies and governments, 'buy' and 'sell' research-related products.[1] It is a place where both researchers and their research are commodified. Researchers gain currency and market position in this place by 'selling' high-value research products such as, for example, peer-reviewed papers in high-impact-factor journals. They can trade this market-based currency (often referred to as their 'research track record') for jobs, promotion, tenure and research grant funding.

At the heart of this research marketplace is competition, which, like the concept of the market, is highly prized, even revered, by exponents of neoliberal thought (Denzin & Giardina, 2017; Cannella & Lincoln, 2015). Researchers can be, and are, ranked against each other in terms of the relative amount of research marketplace currency that they have. This is almost always expressed in the form of research-related metrics which quantify a researcher's research production in terms of market-favoured parameters – number and type of publications, and amount of external research funding gained, being two of the most favoured metrics (Cheek, 2005, 2006, 2011, 2017a, 2017b; Spooner, 2017). The assumption, often so taken for granted or normalised that it does not need to be stated, is that the more marketplace currency the researcher has (that is, the better their metrics), the better the relative worth of that researcher, and *the more credible the research itself.* Putting all this together, we end up with a series of assumptions that have become so normalised as to attain 'truth' status in the research marketplace (Foucault, 1984).

Thus, metrics are used as a type of currency converter in this competitive research marketplace. Metrics convert research outputs – for example, publications and research funding – into the currency of the marketplace, by number, type (journal article, chapter, book and so on), impact of articles published, or amount of dollars gained for research funding, and from which source. These research metrics in turn can be used by a number of stakeholders in the research and higher education arenas to conduct audits, in order to rank the relative research performance of individual researchers, departments within institutions and even institutions themselves (Cheek, 2017a, 2017b). Such stakeholders include managers of researchers and research, and governments who attach funding to performance in those rankings and who make public the rankings of institutions on a range of selected research-related metrics (Cheek, 2005, 2006, 2011, 2017a, 2017b; Spooner 2017). These rankings, in turn, can be re-converted into, or cashed in for, large shares of discretionary performance-based funding from governments, or may attract more full-fee paying students on the basis of being a highly ranked institution on a range of narrow metrics.

At the level of the individual researcher, an unstated assumption in all this currency conversion in the research marketplace is that the 'better' the research metrics attached to an individual researcher, then the 'better' that individual researcher is. The associated amount of metric-derived currency can be used to establish the relative ranking of an individual researcher when compared to other researchers for, say, promotion or tenure.

One effect of this research marketplace phenomenon has been the emergence of our 'data doubles', which, according to Haggerty and Ericson, 'circulate in a host of different centres of calculation and serve as markers for access to resources, services and power... They are also increasingly the objects toward which governmental and marketing practices are directed' (Haggerty & Ericson, 2000, p. 613). The data double – described by Poster as 'the multiplication of the individual, the constitution of an additional self' (Poster, 1990, p. 97) – a form of audited, algorithmically derived self made of data, is so entrenched in contemporary research contexts that it has become taken for granted as ordinary or normal.

Common taken-for-granted data doubles displayed in the research marketplace include the 'h-index-self', the 'number of publications-self', the 'amount of research dollars gained-self' and the 'amount of timely completions of research students that the data double has supervised-self'. Sometimes all these data doubles are put together, perhaps in some form of spreadsheet for managers of research performance to create a type of supersized data double research-self.

Further, as individual researchers, we can *ourselves* convert our research products into market currency. For example, we can create a publication-related profile on Google Scholar, which, among other things, can automatically calculate our h-index[2] and if desired can make this number public – thereby creating a self-constructed and publicly displayed data double. Metrics, such as a researcher's h-index, or Google Scholar profile, have morphed into – and become synonymous with – his or her research track record, which, in effect, is reduced to an auditable statement of the amount of metric-derived currency held by the researcher in their research marketplace bank account. Viewed in this way, track record is a form of data double research-self. Put another way it is simply a commodity in the research marketplace.

The amount of metric-derived currency in that research marketplace bank account is a defining feature of the researcher self and its worth in neoliberal contexts. A data double, for example in the form of a public profile on Google Scholar, can then be used by others and by *ourselves*, to monitor and track our progress on the metrics that are used to construct that profile. How are we going? How do we compare with others who may be our competitors for jobs, promotion or funding in the research marketplace? This can lead to increasing self-surveillance and self-governance by researchers, which may well result in complicity with the metric-driven demands of the research marketplace.

Consequently, researchers increasingly find themselves in research marketplace metric-derived quandaries. For example, where should they publish and why? What data double do they want the publication to contribute to? Is it getting the research itself published, or getting the research into a particular journal that matters most to the researcher? If the honest answer is getting the research into a particular journal, then what is the reason for that? Is it the metrics associated with that journal, such as its impact factor, or could there be some other reason?

The rest of this chapter discusses some of these research marketplace quandaries; how researchers have navigated them and positioned their researcher self in the research marketplace as a result of the chosen navigation path. The hope is that in grounding the discussion in actual situations faced by actual researchers, we can learn from what happened and why, in these instances. This learning can then be used to help us grapple with a fundamental issue in all this – how far to adapt our researcher self, and our research, to the demands of the research marketplace. The next section specifically explores how asking hard and uncomfortable questions of himself led Agewell (2003) to refer to himself as a 'clever idiot', as well as how not asking hard and uncomfortable questions of themselves led participants in Chubb and Watermeyer's (2017) study to lie about, or at least write fictional accounts of the impact of their research.

Interrogating Ourselves: An Antidote to Becoming 'Clever Idiots'

It is not enough to ask ourselves the hard questions. We also have to be ruthlessly honest with ourselves when answering these otherwise often unspoken, and sometimes not even thought-of questions. A way to begin this process can be to ask ourselves when was the last time we actually thought through questions like those posed by Fochler and de Rijcke (2017):

> How do we feel about being lovingly coerced into working long hours, and spending our weekends writing grant proposals and articles, and grading students' exams? Does engaging in the indicator game also change the way we actually do our research work? If so, how does this affect our joy in being an academic, and the way we are able to convey the potential attractiveness of this occupation to our students? Could it be otherwise? Could we then be other, better, versions of ourselves?
>
> *(p. 30)*

Similarly, when did we last consider whether we have simply been too busy or caught up in agendas driven by others, such as our managers, a promotion panel, a publisher or a funding body, to interrogate ourselves about what we are doing and then act based on the answers?

Stefan Agewell did exactly this by asking some hard and very uncomfortable questions of his researcher self. Like many researchers, Agewell served as a peer reviewer for academic journals. Participating in the peer review of other researchers' work is viewed by researchers as making an important contribution to their field of research in terms of ensuring the scientific rigor, development and growth of that field. This is precisely why they undertake this work and donate their time and expertise to it. However, reading about the enormous annual profits being made by commercial academic publishing houses of journals and books, Agewell began asking different questions about his role as a peer reviewer. In a letter published in *The Lancet* journal (2003), he questions whether he is 'a clever idiot' in terms of providing free labour in his leisure time to review scientific papers for scientific journals owned by publishing companies that then make large profits from them.

Continuing to reflect on the commercialisation of academic publication also led him to question whether he is being an even bigger 'clever idiot' when he himself submits a paper to a journal for publication. He spends months planning, executing and writing a paper about a research study. If the paper is accepted by the unpaid peer reviewers 'working' for that publisher, he, as author, often assigns the copyright of the paper to the publisher as a condition of being published. He may even have to pay for publishing costs in some journals. After this, the library of the institution for which he works, and which therefore has paid for his time to do this research and write the paper in the first place, pays the publisher in the form of some sort of journal subscription so that he (and others) can access his own research paper. As he puts it 'Thus, first the institution pays me to do research, then I give away copyright of the results to the publisher, then the library of the institution buys the right to print my paper back from the publisher' (Agewell, 2003, p. 1659).

The letter does not state what Agewell intends to do about this situation – would he change his approach to peer reviewing articles for these commercial journals? This is not the point here. Rather, the point is that it is demonstrable that asking these hard questions of himself, and the actions that he was taking, enabled Agewell to view an otherwise very ordinary taken-for-granted aspect of academic research life in a new light. Such a re-viewing of this practice enabled him to ask new and different questions about both the practice and the role he played in it. Who was he really serving as a peer reviewer? Who actually bene-fits from this, and how? Bringing these types of questions into focus opens up the possibility for Agewell to refuse to become a 'clever idiot' by default. In this instance, surfacing some hard and confronting questions about what he has chosen to do and why, in turn might enable him to make choices about what he is doing and why, and to more fully appreciate the consequences of those actions.

The participants in Chubb and Watermeyer's 2017 study might also usefully have asked whether they themselves were 'clever idiots', but they did not. Chubb and Watermeyer (2017) studied the way academics in the United Kingdom and Australia wrote their applications when applying for funding for their research projects. Their specific focus was the section of a funding applica-tion that requires researchers to identify the 'impact' of the proposed research, or, 'how they will ensure economic and/or societal returns from their research' (p. 2362). Chubb and Watermeyer were interested in what these academics wrote about such perceived returns *and*, of equal importance, why they wrote what they did. They were interested in the thinking behind the writing. In brief, they found that exaggerated Pathways to Impact statements (PIS) were rationalised by interviewees on the basis of 'systemic pressure affecting academic behaviour' in general (e.g. hyper-competitiveness in attracting extramural grant funding; need for 'self-marketability', etc.) and issues related to PIS specifically (e.g. belief that 'everyone' inflates their impact claims) (p. 2364).

The following excerpts from interviews with participants in the study (Chubb & Watermeyer, 2017) provide insights into this thinking behind their writing – why the participants wrote about the impact of their research in the way that they did and/or how they justified that writing:

- 'It might require a bit of imagination, it's not telling lies. It's just maybe being imaginative.' (p. 2367)
- 'If I want to do basic science I have to tell you lies.' (p. 2364)

- 'It's virtually impossible to write one of these grants and be fully frank and honest in what it is you're writing about.' (p. 2364)
- 'I don't think we can be too worried about it. It's survival.... People write fiction all the time, it's just a bit worse.' (p. 2365)
- 'People might, well not lie but I think they'd push the boundaries a bit and maybe exaggerate!' (p. 2368)

The participants in this study used what Ball (2001) calls '"necessary fictions" which rationalise our own intensification or legitimate our involvements in the rituals of [audit] performance' (p. 216). Such necessary fictions in this case include that lying is not lying; everyone does it; I can't survive or be successful in a research marketplace without writing (or lying) like this. In this way the participants justify, even legitimate, the decisions that they took when writing of the impact of their research. Ryan (2012) likens such market-driven and compliant behaviour with the research marketplace to what she terms the 'zombiefication' of academics 'infected by measurement madness, the audit culture, surveillance' (p. 5).

Chubb and Watermeyer's study (2017) reveals a process of *self-zombiefication* in which academics actively comply with a market-preferred brand of research impact (i.e. publishing in high Impact Factor journals; obtaining extramural grant funding; etc.). This self-zombiefication is driven by the desire to be competitive in a research marketplace. It is also driven by a fear of not measuring up, and therefore not being competitive in that place. Thus, the writing of market-friendly, fictitious impact statements is driven by 'the micro-politics of little fears' (Lazzarato, 2009, p. 120) – something not declared and/or acknowledged by the participants in the research.

Put another way, these researchers chose to ignore, or at least not question, effectively lying about aspects of their proposed research in order to increase their chances of getting the grant. Further, as their interview excerpts reveal, they chose not to question why maximising their chances of gaining competitive research funding was more important to them than being honest about the perceived impact their research would have.

The result of all this was a mindless and thoughtless self-driven adherence to market-derived pre-scripted ways of presenting research products when describing the impact of the research that they were seeking funding for. Such mindlessness and thoughtlessness avoids the need to ask uncomfortable questions about what is being written and why. A form of 'functional stupidity' (Alvesson & Spicer, 2016) results where there is 'inability and/or unwillingness to use cognitive and reflective capacities in anything other than narrow and circumspect ways... Functional stupidity means thinking within the box: overadaptation to set ways of thinking and acting' (p. 239). This thinking within the box and process of overadaptation enabled the participants in Chubb and Watermeyer's study to censor their own internal conversations as part of what Alvesson and Spicer refer to as a process of '(s)elf-stupidification' (p. 90). In this process, 'Doubts are cast aside. Critique is culled by the internal censor' (Alvesson & Spicer, 2016, p. 90). What matters most is getting the funding and this matters not only to enable the research to be done, but also to make sure that one's relative research marketplace position is maintained.

The study by Chubb and Watermeyer (2017) and the letter by Agewell (2003), highlight that, as researchers, we make decisions all the time about how, for example, we will or will not do our research, and/or what we will or will not do when writing about our research or participating in the research marketplace. Part of this decision making is related to how far we will, or will not go, and why, in the quest to enhance our metric self and attain valuable research market currency such as grants or publications. As Spooner (2017) reminds us, the obsession with metrics and the audit mentality of the research marketplace is both 'imposed and internalised' (p. 906).

I continue to explore these ideas in the next section of the chapter. Putting the spotlight on my researcher self and my own actions as a vehicle for this discussion, I will explore a quandary in which I found myself recently. This arose when I was thinking about where, how and why to publish the findings of a research study of which I was part.

Grappling with a Hard and (Very) Uncomfortable Question: How Far Am I Prepared to Adapt My Research in Order to Get It Published?

I had the privilege of being part of a multi-disciplinary and multiple methods research program, looking at the way mindfulness was implemented in an elementary school classroom some 20 years ago (Cheek, Lipschitz, Abrams, Vago & Nakamura, 2015; Cheek, Abrams, Lipschitz, Vago & Nakamura, 2017). One of the studies in the multiple methods research program included the qualitative analysis of a complete data set of letters written by the elementary school children about aspects of their experience in undertaking that mindfulness training-based curriculum.

The team of which I was part wanted to get the results of this qualitative study published in a journal that they knew would be read by researchers in the area of mindfulness and education who were not familiar with qualitative research. This was because we wanted to communicate both the study results *and* the potential of the qualitative research approach to those who might not be familiar with, or had little idea about, qualitative inquiry and/or the difference it could make to understandings in this area. Consequently, we decided to submit the manuscript about our study to a journal that was read by the audience we were targeting, even though it had published only a handful of qualitative studies.

When the reviews of the manuscript came back, they were positive, with mostly relatively minor revisions suggested. However, there was one revision required that was anything but minor. The editor wrote:

> Your Results section needs to be tightened considerably. It is more in the nature of interpretation as opposed to presentation of what the students said. I have marked just one paragraph as an example, but the whole section needs considerable work. Omit all discussion – and reference to other researchers – in this section.

The editor went on to say,

> As per the APA[3] Publication Manual (6th ed. P. 32), discussing the implications of the results should be reserved for presentation in the Discussion section. Journal XXX[4] does not allow authors to relate their current findings to previous research publications. Even if I make an exception in this case, the production staff will simply omit them, making your Results section appear disjointed.

This left us with, and in, a dilemma. An expedient, seemingly common-sense response to this request from the editor to present our qualitative inquiry in this way would have been to 'just do it' and thereby to get the paper published. Just doing it would mean pulling apart the rich and textured description of both the research process and the findings into a more traditional style of reporting, much more in keeping with notions of quantitative research. On the other hand, just doing it would also mean getting the paper accepted, and banking the currency that the published paper would give us in a research marketplace.

Thinking about whether to 'just do' what the editor wanted led me to conclude that this would in effect mean that Journal XXX and its editor would be determining the way in which qualitative research can be seen and reported, even if that way violates established qualitative principles, such as the possibility of using other studies if, for example, employing theoretical sampling in both data collection and analysis. I also reached the conclusion that to do what we were being asked to do would mean being complicit in carving up and packaging an iterative non-linear qualitative research process into linear self-contained sections such as Results (and a traditional version of what results are, at that) and then Discussion. I realised that if we did this, it would mean deliberately representing qualitative inquiry in a way that is contrary to my understandings of it in order for the study to be published.

Further, the note from the editor (who is editor on the basis of perceived scientific credibility and standing, which is what enables him to make decisions about the science of the manuscripts submitted) reveals that even he can be overruled by the Journal's production staff as to what gets published and in what form. Using the authority of the APA guidelines, the production staff can determine how the research is reported and have the final say about how the research will be presented. In effect, this hands important decisions about how research, including qualitative research, must be written to the Journal's production staff. They, in turn, derive their authority to make these decisions from their reading of a manual designed to be used to guide (not necessarily to rigidly prescribe) writing style, formatting and referencing.

As a result, the production staff's version of how qualitative inquiry should, even must, be reported becomes normalised in that journal. And, while it does not have a long tradition of publishing qualitative inquiry, there is a risk that such a normalised version could become the standard against which its readers judge qualitative inquiry. This, in its turn, could give rise to another version of what I have referred to elsewhere as an 'expert non-expert' (Cheek 2017b, p. 27) in qualitative inquiry who uses this normalised version of qualitative inquiry as the standard against which to assess the worth of a piece of qualitative inquiry.

Asking questions of myself and my actions (rather than just focusing on the editor or the production staff) enabled me to identify the quandary I was in. This, in turn, enabled me to 'put some pluses (+) in my thinking' (Cheek, 2017a) – as I have called it elsewhere – about that quandary. If the research team of which I was part went along with the requirements of the editor and the production team, then we would get the paper published. + This would give us research market currency. + We would also get the research findings to the audience that we were aiming for. + The trade-off would be that writing up and reporting the qualitative inquiry had to be done in a way that was not in keeping with my understandings of basic principles of qualitative inquiry. + This could lead to the readership of the journal (who were not necessarily familiar with qualitative inquiry given that the journal did not have a strong track record, or hardly one at all, in publishing qualitative inquiry) coming to view this style of presentation as the standard way of presenting qualitative inquiry. + Who gets to say how qualitative inquiry should be reported and what is the authority to say that based on? + There is a price to pay for the expediency of 'just doing' the required changes. + Complicity in allowing the fear of not being published, or the pressure to be published, to compromise the integrity of the way in which qualitative inquiry is represented. + No longer being able to blame everyone else such as the Journal editor, the production team, or the publication metrics we would be so desperately seeking, if we 'just did' the changes.

So, what did the team end up doing? Well, what we did *not* do was change the paper in ways that we did not believe in, just to get a publication metric. Instead, we wrote back to the editor and gave him a brief summary of the principles of theoretical sampling, iterative research design, induction, deduction and abduction, to try to familiarise him further with how these basic pillars of research design can be used in qualitative inquiry. We also found one of the very few articles using qualitative inquiry that had been published by Journal XXX and pointed out to the editor that its writers had employed the same structure in reporting that we had used in writing up our qualitative inquiry. In all of this, we became part of a process of convincing an editor to overturn decisions about fundamental principles of qualitative inquiry being made, at least in part, by the production staff of a journal on production grounds rather than on the basis of the research itself.

Reflecting on this quandary with the benefit of hindsight, I am, for my part, very aware of the importance played by continually asking questions of myself and keeping the spotlight on my own actions when navigating this quandary. Interrogating myself in this way enabled me to actively determine the point beyond which I would not go in terms of changing what I had written about qualitative inquiry. I could, and did, reach a point where I was no longer prepared to adapt either myself, or my research, any further to meet the demands being made on them. It enabled me to stand my ground and to know that I was standing my ground – even if it meant that the paper might be rejected. This enabled me to make a small and considered stand against the 'existence of calculation' (Ball 2003, p. 215) that, as a researcher, I find myself in daily.

As a postscript, the study did end up being published in a journal that reached the type of audience that we wished for it. So, acquiescing to a self-zombiefication process and/or being some form of a 'clever idiot' wasn't necessary to get the paper published after all. This debunks some of the so called 'necessary fictions' that researchers can get caught up by or tell themselves they need to get caught up by in order to survive in a research marketplace.

Conclusion – Where to from Here?

Asking hard and uncomfortable questions of our researcher self enables the possibility for new spaces to emerge from which to write and think against the data double and/or the clever idiot and/or the necessary fiction positions that otherwise we might unthinkingly take up. Interrogating ourselves is a way of thinking deeply about the choices we have open to us when interacting with the research marketplace, and the actions we take in relation to those choices. This type of thinking, and associated action, opens up the possibility to resist and refuse the spaces constructed for us by a metric-driven research marketplace and to embrace other spaces of our own making. For example, it can open up spaces for us to resist writing our qualitative research to fit the requirements of editors of journals on the surface, a seemingly commonsensical or ordinary practice to adopt in a competitive research marketplace. Or it can open up for us the possibility of embracing spaces and actions such as choosing not to write our qualitative inquiry to fit the requirements of funders.

However, such self-interrogation comes with a price. It can be a risky business. It challenges us to re-consider what we are doing and, maybe, have been doing for a long time. It can disturb practices that we have become quite comfortable with, like being happy with the necessary fictions we tell ourselves and weave around our practices to conceal the reality of what lies beneath. It can even cause us to reconsider the worth of what we have been doing.

Consequently, at times, it may well seem easier or more bearable not to ask these questions of ourselves. As Alvesson and Spicer (2016) point out:

> if you persist in asking tough questions, you are likely to cause problems for yourself. You will most likely upset the smooth workings of a group, threaten relations with key people, and disturb existing power structures. Play dumb and the status quo survives, team relationships continue unthreatened. All this allows you to focus on delivering the goods.
>
> *(p. 75)*

This can lead to a strategy of 'sheltering away, waiting for the neoliberal storm to pass' (Ryan, 2012, p. 10) – doing just enough to keep the neoliberal wheels turning while meeting neoliberal requirements. Thus Randell-Moon, Saltmarsh, and Sutherland-Smith (2013) remind us, 'academics are neither completely powerless, nor removed from the impost of a seemingly infectious neo-liberal audit culture on the academy' (p. 56).

Interrogating ourselves about what we are doing and why, in relation to our research, acts as a reality check for us as qualitative inquirers in market-driven competitive times. It puts the focus squarely back on us as individuals, and on how we position both our qualitative inquiry, and ourselves, in the many contexts in which we research daily. It requires us to think *about*, not just how to *do* our qualitative inquiry (Cheek, 2008; Kvale, 1996), and on what we are doing in the research marketplace, not just how we are doing metric wise. This results in us having to confront and address some personal and hard questions such as how we as researchers understand and manage our connections to, and position ourselves in, such a marketplace. It requires us to confront the hard questions such as: How do we adapt (or maybe even over-adapt) to the logic of this place and what is the effect of such adapting on both individual qualitative inquirers and on qualitative inquiry itself (Cheek, 2017a)?

Confronting ourselves and our actions in this way reminds us that in the research marketplace, we 'are not simply victims' (Ball, 2015, p. 259) – we do have choices that we can make. It is important to remember this, as it is very easy to slip into the comfortable and somewhat easy position of critiquing the tenets of neoliberalism and its undoubted effects on shaping our research: *what* we do that research about (topic), *how* we do that research (methods), and *why* we do that research (research market position), while forgetting the part that we ourselves play in shaping that process. As Fochler and De Rijcke (2017) note, often the pervasive impact of indicators ... [specifically, for our purposes, research related metrics] ... is described as a form of coercion through an outside force, neglecting the ways in which academic actors *themselves* are implicated in indicator games and the new strategic possibilities they offer, in institutions, careers and beyond.

(p. 22 – emphasis in the original)

The hope is that interrogating ourselves and our actions as researchers may help to push against, and avoid, what Ryan (2012) has called the 'zombiefication' of research, academics and academic institutions. She identifies 'governance; audit; workload; workforce; and an acquiescent leadership' (p. 3) as sources of such zombiefication. To Ryan's list of sources of zombiefication we can also add acquiescent researchers who do not ask the hard and uncomfortable questions of themselves, instead seeking shelter waiting for the neoliberal storm to pass – perhaps even unaware that there is a neoliberal storm.

This is why self-interrogation is an important exercise to undertake regularly. It can prevent us becoming one of these acquiescent researchers in that it offers the possibility of making visible our role in such indicator/metric-driven games. In so doing it thereby offers us the possibility of refusing (or at least challenging) the multiplicity of forms of research marketplace-derived data doubles and clever idiots constructed for our researcher selves that we might otherwise unthinkingly adopt. In this way, asking these hard and uncomfortable questions of ourselves can be a means of liberating both ourselves and our research from an audit-driven micro-politics of little fears + the numerous data doubles that co-exist with us in the research marketplace + the clever idiot that we otherwise may become.

Notes

1 In a previous version of this chapter, I referred in a shorthanded manner to this research marketplace as researchforsale.com in order to capture how the academic research marketplace is increasingly one that exists in digital/online spaces. However, since that article was published, a business analytics and market research firm has begun using that URL for its online presence.
2 Developed by Hirsch (2005), the h–index reflects both the number of publications a researcher has and the number of citations per publication. He explained the h-index in this way: 'A scientist has index h if h of his or her Np papers have at least h citations each and the other $(Np-h)$ papers have $\leq h$ citations each' (p. 16569). McDonald (2005) explains this further:

> For example a scholar will have an h value of 75 whose 76th paper on the list has been cited 75 or fewer times, but whose 75th paper has more times. Put another way, this scholar has published 75 papers with at least 75 citations each.
>
> *(para. 9)*

3 *Publication Manual of the American Psychological Association* is a style guide that covers writing style and format. It is widely used for publications in scientific and social science journals (APA 2010).
4 The journal has been anonymized.

References

Agewell, S. (2003). Clever idiot? *The Lancet*, 361, 1659.

Alvesson, M. & Spicer, A. (2016). *The stupidity paradox: The power and pitfalls of functional stupidity at work.* London: Profile Books.

American Psychological Association. (2010). *Publication manual of the American Psychological Association* (6th edn). Washington, D.C.: APA.

Ball, S. (2001). Performativities and fabrications in the education economy. In D. Gleeson & C. Husbands (Eds.), *The Performing School: Managing teaching and learning in a performance culture* (pp. 210—226). London: Routledge.

Ball, S. (2003). The teacher's soul and the terrors of performativity. *Journal of Education Policy*, 18(2), 215–228.

Ball, S. (2015). Living in the neo-liberal university. *European Journal of Education*, 50 (3), 258–261.

Cannella, G. S. & Lincoln, Y. S. (2015). Critical qualitative research in global neo-liberalism: Foucault, inquiry and transformative possibilities. In N. K. Denzin & M. D. Giardina (Eds.) *Qualitative inquiry and the politics of research.* (pp. 51–74) Walnut Creek, CA: Left Coast Press.

Cheek, J. (2005). The practice and politics of funded research. In N. K. Denzin & Y. S. Lincoln (Eds.), *The Sage handbook of qualitative research* (3rd edn, pp. 387–410). Thousand Oaks, CA: Sage.

Cheek, J. (2006). The challenge of tailor-made research quality: The RQF in Australia. In N. K. Denzin & M. D. Giardina (Eds.), *Qualitative inquiry and the conservative challenge: Contesting methodological fundamentalism* (pp. 109–126). Walnut Creek, CA: Left Coast Press.

Cheek, J. (2008). Beyond the 'how to': The importance of thinking about, not simply doing, qualitative research. In K. Nielsen, S. Brinkmann. C. Elnholdt, L. Tanggaard, P. Musaeus, & G. Kraft (Eds.), *A qualitative stance: Essays in honor of Steinar Kvale* (pp. 203–214). Aarhus, Denmark: Aarhus University Press.

Cheek, J. (2011). The politics and practices of funding qualitative inquiry. In N. K. Denzin & Y. S. Lincoln (Eds.), *The Sage handbook of qualitative research* (4th edn, pp. 251–268). Thousand Oaks, CA: Sage.

Cheek, J. (2017a). The marketization of research: Implications for qualitative inquiry. In N. K. Denzin & Y. S. Lincoln (Eds.), *The Sage handbook of qualitative research* (5th edn, pp. 322–340). Thousand Oaks, CA: Sage.

Cheek, J. (2017b). Qualitative inquiry, research marketplaces, and neoliberalism: Adding some +s (pluses) to our thinking about the mess in which we find ourselves. In N. K. Denzin & M. D. Giardina (Eds.), *Qualitative inquiry in neoliberal times* (pp. 19–36). New York: Routledge.

Cheek, J. (2017c). The importance of being a selectively stupid academic. Paper presented at the 13th International Congress of Qualitative Inquiry, May 17–20, University of Illinois at Urbana-Champaign, icqi.org.

Cheek, J., Lipschitz, D. L., Abrams, E. M., Vago, D. R., & Nakamura, Y. (2015). Dynamic reflexivity in action: An armchair walkthrough of a qualitatively driven mixed-method and multiple methods study of mindfulness training in schoolchildren. *Qualitative Health Research*, 25(6), 751–762.

Cheek, J., Abrams, E. M., Lipschitz, D. L., Vago, D. R., & Nakamura, Y. (2017). Creating novel school-based education programs to cultivate mindfulness in youth: What the letters told us. *Journal of Child and Family Studies*, 26, 2564.

Chubb, J. & Watermeyer, R. (2017). Artifice or integrity in the marketization of research impact? Investigating the moral economy of (pathways to) impact statements within research funding proposals in the UK and Australia. *Studies in Higher Education*, 42(12), 2360–2372.

Denzin, N. K. & Giardina, M. (2017). Introduction. In N. K. Denzin & M. D. Giardina (Eds.), *Qualitative inquiry in neoliberal times* (pp. 1–16). New York: Routledge.

Fochler, M. & de Rijcke, S. (2017). Implicated in the indicator game? An experimental debate. *Engaging Science, Technology, and Society*, 3, 21–40.

Foucault, M. (1984). The order of discourse. In M. Shapiro (Ed.), *Language and politics* (pp. 108–138). London: Basil Blackwell.

Gildersleeve, R. E. (2017). The lazy academic paper. Paper presented at the 13th International Congress of Qualitative Inquiry, May 17–20, University of Illinois at Urbana-Champaign, icqi.org.

Haggerty, K. D. & Ericson, R. V. (2000). The surveillant assemblage. *The British Journal of Sociology*, 51, 605–622.

Hirsch, J. E. (2005). An index to quantify an individual's scientific research output. *Proceedings of the National Academy of Sciences of the United States of America*, 102(46), 16569–16572.

Koro-Ljungberg, M. & Wells, T. (2017). Method ol o gies … that encounter slowness. Paper presented at the 13th International Congress of Qualitative Inquiry, May 17–20, University of Illinois at Urbana-Champaign, icqi.org.

Kvale, S. (1996). *InterViews: An introduction to qualitative research interviewing*. London: Sage.

Lazzarato, M. (2009). Neoliberalism in action: Inequality, insecurity and the reconstitution of the social. *Theory, Culture & Society*, 26(6), 109–133.

McDonald, K. (2005, November 8). *Physicist proposes new way to rank scientific output*. Retrieved from http://phys.org/news/2005-11-physicist-scientific-output.html.

Poster, M. (1990). Words without things: The mode of information. *October*, 53, 62–77.

Randell-Moon, H., Saltmarsh, S. & Sutherland-Smith, W. (2013). The living dead and the dead living: Contagion and complicity in contemporary universities. In A. Whelan, R. Walker & C. Moore (Eds.), *Zombies in the academy: Living death in higher education* (pp. 53–65). Bristol, UK: Intellect Books.

Ryan, S. (2012). Academic zombies: A failure of resistance or a means of survival? *Australian Universities Review*, 54(2), 3–11.

Spooner, M. (2017). Qualitative research and global audit culture: The politics of productivity, accountability and possibility. In N. K. Denzin and Y. S. Lincoln (Eds.), *The Sage handbook of qualitative research* (5th edn, pp. 894–914). Thousand Oaks, CA: Sage.

Ulmer, J. (2017). Slow correspondences: Writing toward an Ecocene. Paper presented at the 13th International Congress of Qualitative Inquiry, May 17–20, University of Illinois at Urbana-Champaign, icqi.org.

11

CONTESTING ACCOUNTABILITY METRICS IN TROUBLED TIMES THROUGH COMMUNICATIVE METHODOLOGY

Aitor Gómez González

Introduction

The need to publish in journals that have a recognized scientific impact in the academy has become a hurdle that can cause one to lose sight of the quality of the publication itself, focusing instead only on the number of articles published and its position in the main databases.

The publication thereby becomes important for its metrics and not for its contribution to the international scientific community. Instead of looking for a combination of both factors, namely, the possible policy and social impact of the contribution and its scientific impact in terms of where it is published, the quartile of the journal in which the publication will be published is prioritized, and the importance of the contribution is pushed to the background.

The academic career of research staff, especially newcomers, depends on these publications, and for this reason, many contributions are oriented towards those publications of high scientific impact, neglecting the importance of carrying out contributions that denote the social impact of the research work carried out.

In this chapter, I describe how it is possible to simultaneously achieve scientific, policy and social impacts by applying communicative methodology throughout the entire research process. In that sense, I am contributing to overcoming this dependency on metrics and encouraging publications based on social and policy impact: publications that demonstrate real social utility, contributing to improving the lives of the people.

Social Impact of Research

The social impact of research in the social sciences and humanities (SSH) has been questioned in the last several years. The SSH were not presented as a challenge area in the first Horizon 2020 (H2020) draft (instead, it was related only to other research areas).[1] In the end, international entities and researchers initiated a campaign to restore the SSH in the H2020.

The H2020 is directly related to the Europe2020 strategy, which highlights concrete objectives based on the interests and needs of citizens. One of the pillars in the H2020 is social impact: all research projects funded by the European Commission must demonstrate their potential social impact during and after their implementation. This strategy is grounded in the idea of returning scientific knowledge to society. European citizens contribute through their taxes to European research, and they have the right to share in the benefits of these research projects. In some sense, we are returning to the origin of the social sciences, when the main point was to bring science closer to citizens, responding to their needs with a bottom-up approach.

To learn more concretely the real impact of European competitive research, the European Commission launched a call directly related to evaluating the impact of such research. The IMPACT-EV project, coordinated by CREA (Community of Research on Excellence for All), responded to the call and was funded by the European Commission. The IMPACT-EV project evaluated the scientific impact of SSH research and, more specifically, its policy and social impact (Flecha, 2014–2018).

The research work developed in the 7th Framework Program IMPACT-EV has allowed us to establish four parameters to measure the researcher's degree of involvement in a competitive project and the social impact of this involvement, as well as to establish of a key concept directly related to social impact, namely, *research enabling social impact*. These four parameters are *dissemination, transference, social impact* and *social creation*.

Dissemination takes place when institutions, companies, NGOs, and citizens at large get to know about our work (often through outreach activities). In the European research project INCLUD-ED, funded by the European Commission under the 6th Framework Program, a clear example of dissemination was concretely presented at the mid-term development of the project, when representatives of the international mass media, researchers, politicians and representatives of civil society were invited to a media briefing. During this event, the main results of the project were presented, along with results from other competitive projects. Mass media disseminated the main results of the project across Europe.

Transference (political impact) takes place when policymakers, companies, NGOs, or citizens use our research results to plan and carry out their interventions. One of the dimensions of transference is the policy impact of a research project. "Political impact of research occurs when knowledge is transferred, that is, when decision makers and/or social actors employ the published and disseminated results as the basis for their policies and/or actions" (Reale *et al.*, 2017: 3).

Social impact occurs when the results have been published, disseminated, and transferred and lead to an improvement in relation to the objectives prioritized by society. The Higher Education Funding Council for England (HEFCE) defines an impact as an *effect on, change or benefit to the economy, society, culture, public policy or services, health, the environment or quality of life, beyond academia* (HEFCE, 2016: 4). In a broader sense, the National Science Foundation defines impact as *the potential to benefit society and contribute to the achievement of specific, desired societal outcomes.*[2]

Social creation is one of the concepts that could help in diminishing the importance given to accountability metrics. Social creation occurs when a scientific article is published about a new reality created by its authors. In this way, the main focus is transferred to the creative work carried out by the author himself. It is no longer a matter of publishing about existing elements or realities but of making a social impact with our research, which then leads us to quality publications that can be made available to the citizenship. Just as there is a growing interest in carrying out research that has a social impact, which has a positive impact on citizenship, there will also be interest in publishing articles based on social creation.

> In certain sciences, it is also clear that the value of an article is not in its draft but rather in the discovery that explains, for example, the article a year ago that announced the obtaining of embryonic stem cells from an adult cell in humans. The social sciences require the concept of social creation, which clearly reveals the contributions that are being made to improve the social reality.[3]
>
> *(Flecha, 2014, Message 1)*

One of the main goals of social creation is to improve societies in a way that has not previously been attempted. However, the need to involve science with the social reality it studies, giving it the ability to develop work in common, much more transformative ways, is clear (Aiello & Joanpere, 2014).

Research Enabling Social Impact (RESI)

RESI is the cumulative impact of the many research studies that have contributed to further research that has in turn achieved social impact. To find solutions to problems facing society, there are many research projects and teams exploring different avenues. This multifaceted approach has been necessary to find solutions, facing both failures and successes along the way.

There are different indicators of social RESI that help us to determine whether a scientific contribution could be understood as a RESI. One of these indicators is the collaboration of the contribution's researchers with other researchers who have been successful in achieving social impact. A second element is the establishment of research hypotheses that link possible findings to future societal improvements. Another element is related to publications: if we publish research findings with reference to future contributions to societal goals in scientific journals, we are providing data directly related to a possible contribution based on social impact. In the same way, scientific articles about research cited in articles about new discoveries or findings with social impact are a possible indicator of RESI.

Finally, other indicators of RESI are based on citations in two ways: those related to scientific articles and those related to social media. These two types of citations include citations of a research finding in other scientific publications (i.e., a finding that is cited many times is more likely to become prior knowledge to a finding that leads to social impact) and citations of a research finding in different social media.

As an example of RESI, social impact and social creation, we introduce the research work developed during the INCLUD-ED project. The main objective of this project was to analyze educational strategies that contribute to social cohesion and educational strategies that lead to social exclusion in the European Union. One of the main results extracted from the project was the identification of successful educational actions (SEAs) and how these could be transferable to any other social context.

INCLUD-ED contributed to four official social goals established by the European Commission: (1) increasing employment of the population aged 20–64 (EU target = 75% employed); (2) reducing the rate of early school leavers (EU target = below 10%); (3) increasing the rate of 30–34-year-olds completing third-level education (EU target = at least 40%); and (4) reducing the number of people in or at risk of poverty and social exclusion (EU target = 20 million reduction).[4]

The contribution of these four social goals indicates that the project attained social impact directly related to the reduction of absenteeism and early school leaving, the increase in school performance and the expansion in the number of schools implementing SEAs. Finally, once the project was finished, the SEAs were transferred to other contexts in Europe and Latin America with great success. Such success indicates that the project also attained RESI, being a key point in future successful developments in education.

Communicative Methodology and Social Impact Through the INCLUD-ED Project

CREA has been applying the communicative methodology of research for more than 25 years. The communicative approach has allowed researchers to obtain the results of research oriented to social transformation and to the improvement of life conditions of the groups involved in the research. In short, the research projects carried out with this methodology have had a social impact.

Right now, the H2020 and the new Ninth Framework Program have highlighted the importance of cocreation when researching. That is, the key point is how we can interpret and construct knowledge in a collaborative way, arriving at a consensus among all participants in the research process. In that sense, a communicative methodology has been implemented to achieve this goal for the past two decades. The research process and the results of any research project using this methodology offer clear examples of how we can involve participants during research and how we can attain social impact precisely by working together with the potential participants of a research project. In this section, we introduce how social impact was attained through the INCLUD-ED research project (Flecha, 2006–2011).

Classic books on research methods (Bogdan & Knopp Biklen, 1998; Creswell, 1998; Denzin & Lincoln, 2015; Milinki, 1999) state that research always starts by identifying a research problem. According to this previous literature, setting the research problem involves the important effort of reading the existing research literature in order to identify the "gaps" within the previous studies to focus on when designing the new study. However, in Europe, it is increasingly common for research agencies to design their own research agendas aligned with the demands emerging from the citizenry through policy agendas. The authorities define the research topics, and then researchers meet in multinational and multidisciplinary teams (partnerships) to address these topics. In this sense, the challenge is not simply to find "literature gaps" but to establish research structures (networks, partnerships, etc.) through which to collaborate with different research teams and institutes across Europe.

Hence, the *classic* research design is redefined in those terms, giving rise to new research challenges around the need to collaborate among different researchers, from different sites, with different local contexts, and encouraging the interaction between these researchers and the potential participants in a research project. That was a key point in INCLUD-ED. The purpose of the study and the research questions were not open for any discussion since they came from the funding authority; the European Commission, through the *open call for research projects*, clearly established the societal challenges researchers must address.

INCLUD-ED illustrates contemporary new challenges for researchers in a globalized and interconnected world in order to attain social and policy impact. These challenges include both technical issues (related to the research design itself) and more holistic components (emerging from the '*objects*' being studied). Figure 11.1 summarizes these types of challenges and how to use communicative methodology if we desire to attain real social and policy impact with our research.

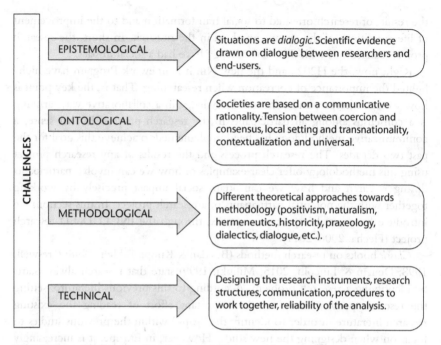

FIGURE 11.1 Challenges of attaining social impact using a communicative approach

Source: own creation.

Epistemological

Gómez et al. (2006) claim that our *object of study* is defined as being "dialogic." The international scientific community has discussed for a number of years whether *evidence* is objective (positivism and neo-positivism), constructed by researchers (social constructivism), or the consequence of a dialectic relationship between all the subjects involved (sociocritical approaches). The most recent analysis suggests that social phenomena are complex objects of study in which individuals actively participate, and hence, we need methodologies to cope with this type of situation.

The communicative methodology incorporates the voices of the end-users even within the process of deciding and selecting the forms of data and the instruments with which to collect them. This process is accomplished through the establishment of an advisory committee formed by end-users that meets with the research team several times before and during the study to make sure that the data selected, collected and used (a) are the best with which to answer the research question (which emerges from the request that citizens have asked policymakers to address through the public research agenda); (b) meet all ethical rules to avoid any misuse of data (and that the research does not use data for other purposes than those established in the agenda); and (c) produce social

TABLE 11.1 Communicative Organization of the Research in the INCLUD-ED Project

Body of Research	Aim	Participants
Advisory Committee (AC)	The AC is a consultancy body containing a wide range of representatives of vulnerable groups. The purpose of this committee is to guarantee the presence of the vulnerable groups in the project and that the consortium provides recommendations that are meaningful to them.	Members from NGOs including an organization working with blind people, an association of participants in adult education, an Arab and Muslim socio-cultural association, a cultural association of Romanian people, volunteers in schools, members from foundations working with migrants or ethnic minorities, members of multicultural groups, members of women's organizations.

Source: own creation.

impact in the terms proposed by Reale *et al.* (2017): they serve to improve the end-users' situation.

Ontological

Habermas (1984) studied the nature of human action in terms of rationality. According to him, the classic theory of rationality proposes that individuals may choose between instrumental (teleological) rationality or authoritarian rationality (based on the use of force due to a position of power). However, he stated that there is another way to justify and explain our actions, which is *communicative* rationality, meaning the kind of situation in which individuals use the force of the best argument to determine their actions. This approach means that "reality" (or the world) is not just a result of dialectical relationships between social, cultural, ethnic or economic groups (such as migrants vs. natives, working class vs. capitalists, slaves vs. free men and women, etc.); instead, our world is a human construction whose meanings are constructed in a communicative way through the interaction between people. Drawing on this ontological assumption, the communicative methodology analyzes how to reach consensus around certain actions in a communicative way.

If social phenomena emerge from communicative situations in which participants use consensus, coercion, and instrumentality (in different combinations), then we need to figure out how to identify the components (behaviors, norms, procedures, etc.) aligned with the aim of the study. The INCLUD-ED project was aimed at studying the communities involved in learning projects that contribute to reducing inequalities and marginalization.

To achieve this goal, the members of the research team created a research organization (an advisory committee, as we have seen previously). The practical strategy used to identify the components oriented towards accomplishing this goal was the development of codes with which to analyze the data collected using the *transformative* and *exclusionary* dimensions, i.e., considering the aspects that would lead to the project's achievement of its aims vs. the aspects that would make that achievement difficult or would keep it from occurring. This taxonomy (transformative vs. exclusionary dimensions) is one of the main features of the communicative methodology. Table 11.2 illustrates the codes developed in INCLUD-ED to analyze the data generated within the research project.

The codes used in this table were developed a priori in accordance with the INCLUD-ED research questions. This procedure is also used when researchers draw on their literature review to create codes consistent with the findings already identified and accepted by the scientific international community. However, this is not the only way to define codes of analysis. We can also develop them a posteriori, based on the data provided by the participants in the study. Usually, the researcher reads the transcripts of the data collected, identifying key themes that become categories of analysis. The process ends when the taxonomy is *saturated* (that is, when no new information arises from the data).

In both cases, the communicative methodology always organizes the codes between transformative and exclusionary components. The research team must also reach consensus on the categories of analysis, using valid arguments to justify each of the categories proposed. INCLUD-ED organized several meetings with all the participants in the project to verify and agree on the categories included in the grid of analysis.

Methodological

This set of challenges is interwoven with the previously discussed challenges, since the methodological approach chosen by the researchers depends on their epistemological and ontological assumptions. In this sense, a study may be mainly quantitative when researchers want to conduct an "objective" study drawing on the idea that our societies are *social facts* (in Durkheim's terms) that are susceptible to objectification (that is, we can measure them, count them, and describe them without modifying them with our work). Drawing on this approach, researchers use experimental (*uni-* or *multi*factorial analysis) or quasi-experimental (nonequivalent groups, temporary series, etc.) designs (Kim & Mueller, 1978; Ward & Gleditsch, 2008).

However, other authors assume that social phenomena result from subjective construction. A *classic* reading in this perspective is Berger and Luckmann (1966), *The Social Construction of Reality*. In this book, the authors illustrate how human beings create our society as a result of decision-making processes.

TABLE 11.2 Analytical Grid for Project 6 (INCLUD-ED)

	Transformation of the Neighborhood				Transformation in the Lives of the Participants			
	Housing	Health	Employment	Social and Political Participation	Lifelong Learning	Empowerment	Social Cohesion
Exclusionary dimension	1	3	5	7	9	11	13
Transformative dimension	2	4	6	8	10	12	14

Source: extracted from Gómez, Latorre, Sánchez and Flecha, 2006.

First, we decide how to organize our everyday world; then, every decision becomes institutionalized as time goes on. Berger and Luckmann use the example of an isolated island to explain this process. Two people (a man and a woman) survive a shipwreck. They end up on an island. No one else lives there. They begin to organize themselves. One goes to hunt, whereas the other one stays "at home." All tasks are divided among them. Some years later, they have children. The children learn their parents' society. For them, the division of tasks is something that already exists as it is. This is what we call "institutionalization," and socialization is how we teach newcomers how things are and how they function. Drawing on these assumptions, constructivist researchers tend to use qualitative methods to analyze and interpret social phenomena. The methodological challenge is to find techniques that may help the researcher to understand the topic of study, assuming that scientific evidence is a social construction. There are several modalities of research that are well accepted internationally, including ethnographies, phenomenology, narratives, biographies, case studies, etc.

Some researchers also include the idea of "power" in the picture. According to them, every single social fact is the result of a relationship in which one or a few participants have more power than others. Social phenomena are the result of dialectical relationships between individuals holding unequal power positions. Most of the studies in the field of *history* adopt this framework (so-called *historicity*). A well-known example would be Marxist studies. Researchers take a sociocritical position towards data. They look for praxeologies to explain how things work. The modalities of investigation include action research, participative research, etc.

Sociocritical, constructivist or objectivist research designs have been widely used in social sciences and humanities research domains (Gómez et al. 2006). There is a large body of literature illustrating and explaining how to deal with research challenges drawing on those methodological approaches.

However, the rise of dialogue in our societies (Flecha, Gómez & Puigvert, 2003) introduces a new set of challenges with which none of those methodological approaches are equipped to deal. We need to include *dialogue, consensus, justification*, and *interaction* among our categories of analysis. These words refer to a complex reality in which participants have other sources of action in addition to the *traditional* ones (meaning power or teleological rationality). Are these categories universal? Is this methodological approach valid and useful to conducting research in any type of social setting? Can we draw on it to work collaboratively with researchers from different regions in Europe, with different local contexts?

Technical

What type of instruments do we need to use? How should ethics be dealt with in different contexts? How does a communicative methodology provide procedures and criteria to attain social impact? If the previous challenges suggest that social phenomena present some commonalities that we can use in order to overcome the problem of contextualization (specificity of the local setting), the last challenge deals with the fact that instruments must be useful to work in different locations, implemented by researchers with different theoretical and methodological backgrounds, with people using different languages and different cultural references; all of these are factors that make it very difficult to cope with a common set of research instruments. *Dialogue* becomes a key factor in overcoming all these difficulties.

When using communicative methodology, it is particularly relevant for all members of the research team to become acquainted with the postulates and theoretical foundations on which this approach is grounded. Its interdisciplinary theoretical bases include Habermas's theory of communicative action (1984), which argues that there is no hierarchy between the interpretations of the researcher and the subject, and their relationship should be based on the arguments they provide and not on their social or academic position. Communicative methodology also draws from Mead's symbolic interactionism (Mead, 1934), which stresses that interactions change people's interpretations and therefore do not depend only on the individual subject. Finally, Garfinkel's ethnomethodological framework (Garfinkel, 1967) is also taken into account for a better understanding of the subjects' insights into their contexts. These theoretical foundations align with the transformative paradigm, which provides a framework for addressing and overcoming educational and social inequalities and injustice (Mertens, 2007).

Keeping these two aspects in mind (universal capacity to argue and importance of the interactions), we used virtual and face-to-face meetings to design the research instruments to collect the data. For instance, we used a questionnaire to investigate how educational exclusion affects diverse sectors of society, particularly the most vulnerable groups (i.e., women, youth, migrants, cultural groups and people with disabilities) and what kind of educational provision contributes to overcoming their respective discrimination. The researchers used previous questionnaires already validated internationally as a starting point. We selected batteries of questions and produced several drafts. Each one was discussed by the different research bodies. In doing so, we included the voices of vulnerable groups in the design of the instrument used to collect the data.

The same procedure applies for the communicative daily life stories and the focus groups used during the research fieldwork. Questions, themes, how to address a particular topic, etc., were discussed until reaching an agreement within the research bodies. The review of previous literature was always the starting point to produce the first draft. But that was just a temporary document to support the discussions and create dialogically the final version of each instrument.

This process is highly relevant since sometimes researchers must work with cultural or ethnic minorities with specific cultural rules, and thus, not all standard questions would be welcome in every single community. It was important to include members of the different communities within the research team to obtain feedback from the participants, because it is also true that a researcher or a technical assistant belonging to the cultural group participating in the study facilitates the process of accessing the data.

However, the most important factor in working with diverse community groups was creating spaces of egalitarian dialogue (in the terms of Flecha, 2000–). Participants, especially those coming from vulnerable groups, are tired of researchers leaving behind the contexts studied and the persons participating in the research after obtaining their data. Social impact is an important ethical consideration when working with people living in vulnerable situations.

Concluding Remarks

The four challenges discussed above summarize the most common difficulties that researchers encounter when working collaboratively in attempts to attain social and policy impact. The type of research objects and phenomena that social sciences and humanities researchers must face are complex situations in local settings with specific cultural and social constraints.

Dealing with such complexity demands serious and rigorous but also inclusive and participatory methodological frameworks. Communicative methodology provides tools, such as dialogue, inclusion of everyone's voices, dismantling of the interpretative hierarchy, universality of speech acts, and communicative actions, that allow researchers to conduct research designs that successfully address most of those challenges. Using the communicative organization of the research, creating different research bodies integrating researchers, stakeholders and participants (end-users), and including egalitarian dialogue premises in our analysis help to overcome the challenges of working in an international and diverse environment with research teams from all over the world.

The high complexity of social phenomena in our globalized and interconnected societies raises important difficulties for researchers in finding universal and transferable evidence addressing their research questions. In addition, a trend towards conducting research with social impact is becoming increasingly common across Europe. Our responsibility as researchers is to produce evidence and reliable data so that policymakers have input with which to make decisions to fulfil their political agendas. Local and cultural differences between countries, regions, cities, communities, etc., make our work even harder, since it is difficult to discern how to disconnect the particular from the universal.

The INCLUD-ED project is a clear example of how applying communicative research during the entire research project enables us to attain social and policy impact. All publications derived from the project have had a powerful scientific impact and have, at the same time, shown the social and policy impact attained during the research project. Many articles in prestigious international journals have shown the scientific impact of the project, but the most important fact was the social and policy impact of the research results obtained using communicative methodology and how the lives of the people improved during the research process.

Notes

1 The H2020 (Horizon 2020) is the new European Framework Program on Research that follows principles and recommendations established through the EU2020 (European Union 2020).
2 Retrieved from: www.nsf.gov/od/oia/publications/Broader_Impacts.pdf.
3 Own translation from the original in Spanish.
4 Retrieved from: https://ec.europa.eu/info/business-economy-euro/economic-and-fiscal-policy-coordination/eu-economic-governance-monitoring-prevention-correction/european-semester/framework/europe-2020-strategy_en.

References

Aiello, E. & Joanpere, M. (2014). Social creation. A new concept for social sciences and humanities. *International and Multidisciplinary Journal of Social Sciences*, 3(3), 297–313. doi: 10.4471/rimcis.2014.41.

Berger, P. & Luckmann, T. (1966). *The social construction of reality*. New York: Anchor Books.

Bogdan, R. C. & Biklen, S. K. (1998). *Qualitative research for education: An introduction to theory and methods*. Boston: Bacon.

Creswell, J. W. (1998). *Qualitative research and research design: Choosing among five traditions*. Thousand Oaks, CA, London, New Delhi: SAGE.

Denzin, N. K., & Lincoln, Y. S. (Eds.) (2015). *The SAGE handbook of qualitative research*. Thousand Oaks, CA, London, New Delhi: SAGE.

Flecha, R., Gómez, J., & Puigvert, L. (2003). *Contemporary sociological theory*. New York: Peter Lang.

Flecha, R. (2014–2018). *IMPACT-EV: Evaluating the impact and outcomes of European SSH research* (7th Framework Program, Socio-economic Sciences and Humanities. Project N° 613202). Brussels: Directorate General for Research & Innovation, European Commission.

Flecha, R. (2006–2011). *INCLUD-ED: Strategies for inclusion and social cohesion in Europe from education* (6th Framework Program, Citizens and Governance in a Knowledge-based Society, CIT4-CT2006028603). Brussels: Directorate General for Research, European Commission.

Flecha, R. (2014, May 17). [General debate]. El concepto de Creación Social y evaluación Ciencias Sociales y Humanidades. [Message 1]. [Online forum comment]. Retrieved from http://amieedu.org/debate/index.php?action=profile;area=showposts; sa=topics;u=11.

Flecha, R. (2000). *Sharing words: Theory and practice of dialogic learning*. Lanham, MD: Rowman & Littlefield.

Garfinkel, H. (1967). *Studies in ethnomethodology*. Englewood Cliffs, NJ.: Prentice-Hall, Inc.

Gómez, J., Latorre, A., Sánchez, M. & Flecha, R. (2006). *Metodología Comunicativa Crítica*. Barcelona: El Roure.

Habermas, J. (1984). *The theory of communicative action. Reason and the Rationalization of Society*. Boston: Beacon.

Higher Education Funding Council for England (HEFCE). (2016). *Publication patterns in research underpinning impact in REF2014*. London: HEFCE.

Kim, J. O. & Mueller, C. W. (1978). *Factor analysis: Statistical methods and practical issues*. Thousand Oaks, CA, London, New Delhi: SAGE.

Mead, G. H. (1934). *Mind, self and society*. Chicago: University of Chicago Press.

Mertens, D. M. (2007). Transformative paradigm: Mixed methods and social justice. *Journal of Mixed Methods Research*, 1(3), 212–225.

Milinki, A. K. (1999). *Cases in qualitative research*. Los Angeles, CA: Pyrczak.

Reale, E., Avramov, D., Canhial, K., Donovan, C., Flecha, R., Holm, P., Larkin, C., Lepori, B., Mosoni-Fried, J., Oliver, E. & Primeri, E. (2017). A review of literature on evaluating the scientific, social and political impact of social sciences and humanities research. *Research Evaluation*. Doi: https://doi.org/10.1093/reseval/rvx025.

Ward, M. D. & Gleditsch, K. S. (2008). *Spatial regression models (Vol. 155)*. Thousand Oaks, CA, London, New Delhi: SAGE.

12

SEDUCTION AND DESIRE

The Power of Spectacle

Bronwyn Davies

Since January 2017, we have been witness to an extraordinary *spectacle*. Courtesy of the e-media we can get up each morning to gaze aghast at the latest episode. This spectacle, I will argue here, is the culmination of neoliberal ideology, its rapacious version of capitalism and its systematic undermining of democracy. U.S. President Donald Trump is neoliberalism's ideal-typical product and its avid proponent; he provides a vivid manifestation of its ultimate weakness and with luck its imminent collapse. I may be being too optimistic on the imminent collapse front; I thought the global financial crisis heralded that collapse, but neoliberalism by then was the only discourse that leaders of capitalist countries had access to on both right and left. One of our tasks as qualitative researchers is to explore the entangled agencies that have made this spectacle possible, including the agencies of individual human lives, and the agencies of the assemblages and flows through which the spectacle is constituted. Our qualitative and post-qualitative methods have persistently run counter to the hallmarks of neoliberal control that have been aimed at closing down thought and silencing critique.

Over the last four decades we have worked toward new understandings of the multiplicity and fluidity of humanity, expanding our capacity to express that humanity, not limited by liberal or neoliberal individualisms, rationalities, and prejudices. We have more recently opened ourselves up to our inter-relatedness to, and with, the material universe—and with that opening, developed a greater responsibility (and response-ability) toward the more-than-human universe.

In this chapter, I sketch the contrary assemblages of neoliberalism and the Trump spectacle on the one hand, and on the other, the creative evolution of qualitative research with its commitment to social justice and to communities, both human and nonhuman, that are relational, plural, and always emergent, and in which "each of us is a multiplicity in connection with other multiplicities, even where those multiplicities contain, as they inevitably do, opposing lines of force" (Davies, 2014, p. 9).

Thousands of pages attempting to make sense of the Trump spectacle have already appeared in the news media, not just sensationalizing the latest episode, but trying to work out where on earth it is coming from. Kim and Yancy, for example, in *The Stone*, in an imagined letter to Kim Jong-un, distinguish two competing strands in American culture; one that is diplomatic, peaceful and caring, and the other, which Trump represents, of arrogance, belligerence, bullying and insult. Trump represents, they say, "a strand and strain of the American experiment that stubbornly holds on to the misguided notion that we are a nation of destiny and superiority, strengthened on legacies of white supremacy and rapacious capitalism" (Kim & Yancy, 2017, np), a strand dedicated to expanding the wealth, and power, of a very small minority.

Until Trump was elected, I had failed to grasp neoliberalism's links with white male supremacy. Even though I had researched neoliberalism's impact on academic work, and even though I had left the neoliberalized academy in order to be able to continue to do my work, and even though I witnessed the potentially devastating attack on the work of qualitative researchers wrought by "No Child Left Behind," I had not fully grasped the power of it to undo our work— and not only our work, but the work of decades devoted to creating a less cruel and divided world.

The fear of that work being undone is felt across the arts and social sciences and among activists and entertainers. Faulkner writes, for example, of Mavis Staples' new album, *If All I Was Was Black*, quoting Staples who said:

> "Sometimes I feel like I'm living in the 60s" … "It's like starting all over again…. Every song in this album is about what is happening in the world today, and mainly what's happening in our White House, because I feel that this president that we have is the instigator for bigotry and hate." Particularly abhorrent to Staples is the re-emergence of white power rhetoric in the conservative mainstream "These men were marching through this city, Charlottesville, [she said] and they had torches, the way the Ku Klux Klan used to do, and the only thing different was they didn't have the white sheets over their heads, you know?" … In 1965 [the Staples Singers were] closely associated with the burgeoning civil rights movement and their songs became anthems for the freedom marches being led by their new-found friend and lifelong inspiration Dr Martin Luther King.
>
> *(Faulkner, 2017, p. 20)*

Qualitative research has its roots in the freedom marches, in the belief that social change is both desirable and possible, and that creative evolution or *differenciation* can take place when history is genuinely open to the future (Massey, 2005).

Trump's notoriety as a white male supremacist has long since been established. In 1989 when a young white woman was raped in Central Park, five teenagers, four black and one Latino, confessed following police torture, to having been involved in the rape. Trump took out full-page ads in the major New York papers, calling for the return of the death penalty. When the five boys were later cleared on the basis of DNA evidence Trump continued to ride the hobby horse of their black guilt and to call for their lynching. His fanatical hatred of the innocent boys resonated with Ku Klux Klan narratives of white womanhood defiled by dark savages—a world with links to the hierarchical, vengeful, arbitrary and rule-bound discourse of the Old Testament.

The Spectacle

The Trump spectacle has provided us with wonderfully rich data to work with. The video clips of Trump, and of his Greek chorus cheering him on; his tweets; the protesters; the comedians; the political activists; the judges; the journalists of the alt-right and those holding the ground of critique. We can draw on this richness of data to explore the spectacle, and our entanglement in it; we can analyze its seductive force—even those of us who hate it are compelled to watch it—and we can perhaps come to new insights about the significance of our work.

Trump the President, the seducer/trickster/Pussy-Grabber-in-Chief (as some Australian journalists call him), is his own most avid audience. He is both seducer and seduced:

> The shows [the President watches] and the President are stuck in a bizarre feedback loop. There are televisions wherever he goes—Trump Tower, Mar-a-Lago, the White House. It's an eerie, modern version of the pond of Narcissus.
>
> This is our new reality. The leader of the free world is a fabulist addicted to the glow of his own image.
>
> (McKenzie-Murray, 2017, p. 11)

That reflected image needs a great deal of avid attention—the pond's surface must, if he is to be happy, reflect back at him an image of undying love.

At a recent meeting of leaders in the Asia Pacific, Trump stood in the lineup of presidents and prime ministers, all dressed in identical light white cotton suits provided by the hosts in the Philippines—an ASEAN (Association of Southeast Asian Nations) tradition, which invariably makes the assembled leaders look a little foolish—no longer quite in control of their image. Part of their challenge is to maintain their dignity, and their sense of humor, while they walk out onto the stage in the outfits that are each year somehow typical of the host nation's mode of dress. In November 2017, Trump stood in the middle of the line, a positioning he has, in the past, been caught on camera rudely shoving other leaders aside for. He towers over the others. While they stand still, looking forward, waiting patiently for the camera to do its work, Trump keeps looking around at the others, apparently disconcerted that they're not paying attention to him. He starts rhythmically swaggering like a schoolyard bully, vulnerable, aggressive, pitiful, dangerous. The others stand still, gazing at the camera. Trump continues to swagger, and to look about him for approval.

Seduction

Thirty years earlier, Trump (or his ghost writer) in *The Art of the Deal* described his philosophy as one of self-promoting bravado, through which he persuades others to want the fantasy he provides them with:

> The final key to the way I promote is bravado, I play to people's fantasies. People may not always think big themselves, but they can still get very excited by those who do. That's why a little hyperbole never hurts. People want to believe that something is the biggest and the greatest and the most spectacular. I call it truthful hyperbole. It's an innocent form of exaggeration—and a very effective form of promotion.
>
> *(Trump & Schwartz, 1987)*

Jean Baudrillard, writing in 1990, three years after *The Art of the Deal*, said, "The capacity imminent to seduction [is] to deny things their truth and turn it into a game, the pure play of appearances, and thereby foil all systems of power and meaning with a mere turn of the hand" (1990, p. 8). For seduction to succeed it must enchant, and enchantment, Baudrillard says, "begins only after one has been taken in by one's desire" (Baudrillard, 1990, p. 70). Our desire makes us active in our own seduction.

As if following Baudrillard's advice, the Trump machine in 2016 used social media to extract the information that would enable him to tailor his hyperbole to fit and amplify the fears and desires of his various audiences. He told them government could not be trusted; the truth was unavailable as no-one could be trusted, least of all journalists; there was a malaise in the country that could only be addressed through closing the borders and expelling foreigners, and through tax breaks for the rich. The feedback loop of desire became an echo chamber, where the basest fears and desires of his many admirers were magnified. They were seduced by his apparent capacity to speak directly to them. As Baudrillard (1990, p. 68) points out:

> What seduces is ... the fact that it is directed at you. It is seductive to be seduced, and consequently, it is being seduced that is seductive. In other words, the being seduced finds himself in the person seducing. What

the person seduced sees in the one who seduces him, the unique object of his fascination, is his own seductive, charming self, his lovable self-image.

Both Trump and his supporters, in this analysis, find themselves in each other, simultaneously producing themselves in each other's image. The watery ponds in which they gaze at their own image are filled with the rank weeds of the neoliberal assemblage.

Those rank weeds are characterized by a shift to making money as the sole value and the de-valuing of critical thought. Neoliberalism has been intent on shutting down critical thought since the 1970s. School education has tended to become more practical, and academic disciplines focused on critique, such as women's studies, history and philosophy, have had their funding reduced and in many places have closed down. The simultaneous burgeoning of social media has played into this trend to the extent that absurd misinformation about Hillary Clinton could be read as true. Trump's trumpeting of "fake news" has been accompanied by, according to fact checkers, an average of five pieces of misinformation a day. At the same time, he is heard as speaking truths that establishment politicians are not willing to utter. Even if he is wrong, it is refreshing, for many, that he says what he thinks. So, although many of us gaze at him with horror, he is a curious manifestation of social forces at play in the capitalist world that his presidency draws our attention to. Many who work closely with Trump see his behavior as problematic (Haberman, Thrush & Baker, 2017), but almost half the country voted for him, and many of those supporters still fervently believe in him. It is interesting to ponder the nature of his seduction of those supporters.

The sorcery of seduction, Baudrillard tells us, is found precisely in its nonsense:

> against all logic, it is the improbable prophecies that come true; all that is required is that they not make too much sense. Otherwise they would not be prophecies. Such is the bewitchment of magical speech, such is the sorcery of seduction.
>
> *(Baudrillard, 1990, p. 75)*

He goes on: "Any system that is totally complicit in its own absorption, such that signs no longer make sense, will exercise a remarkable power of fascination" (Baudrillard, 1990, p. 77).

One of the side effects of seduction, working as it does through the symbolic and against rationality, is the power of the seduced to ignore those slippages that risk breaking the magic spell. Don Juan, for example, the archetypal seducer, could trust his audiences to blot out his failures. The best romantic heroes are, after all, flawed; to love them is to forgive those flaws, to engage, in Clément's words, in "mystical forgiveness, pursuing the beloved with charitable parables" (Clément, 1989, p. 94).

And Trump's flaws are legion. He pushes and shoves, he swaggers, and he threatens to drop the "mother of all bombs"; he actively pursues policies that dangerously escalate global warming, which will cause sea-level rises producing a refugee crisis far greater than the one we have at present. He is a dangerous bully-boy on the international stage. But he is by no means the only bully-boy occupying a presidency and supported by his people. He may be less polished than some, and less willing to be groomed by his advisors than they might wish. But isn't that something we have come to dislike about so many of today's politicians, that they only speak the lines they have been given, and are afraid to risk deviating from their scripts? The interpretations of his utterances, written and spoken, differ wildly, from "he is just plain stupid," to a clever set of ploys that actually have behind them some kind of guiding philosophy. But what might that philosophy be?

Slippages or a Coherent Philosophy?

The men Trump says he admires are ruthless dictators, for whom exterminating undesirables is everyday work that they boast of. He said at the ASEAN summit meeting in Manila in November 2017, for example, that he had a "great relationship" with President Rodrigo Duterte of the Philippines. Locally, the men he says are "great guys" include Roy Moore, who is said to molest young girls and who believes Muslims shouldn't serve in Congress, that homosexuality should be criminalized, and that the Bible should replace the law. Trump boasts of dominating women through sexually assaulting them and has started to dismantle women's rights to control their fertility, causing an inevitable rise in female mortality. Like Duterte and other brutal dictators, he has created classes of people who can be dominated, and others who can be eradicated.

In the mirror reflection back and forth, in the cycle of seduction and power, who catches these images and reflects them back? No one is quite sure how they add up, except for his admirers, who probably understand them as neatly summed up in the slogan "make America great again." I was, in the beginning, mystified by the power of that slogan. Wasn't America already the most powerful nation on earth? I was yet to discover the particular nature of that greatness to come, with its roots firmly in white male supremacy and in rapacious, unprincipled global capitalism, a greatness that requires the undermining of democracy, repeated displays of power over other nations, and over individuals and groups within, and in the final event, power over and potential destruction of the planet itself.

How Has This Seduction Worked?

The seduced, caught in the feedback loop, do not believe their seducer will fall; they are entangled with him, finding themselves mirrored in him in a mutual seduction in which he cannot be allowed to fail. His power rests in very large part in the continuing success of his seduction: "If power cannot be 'exchanged' in accord with this minor cycle of seduction, challenge and ruse, then it quite simply disappears" (Baudrillard, 1990, p. 45). Without seduction, there is no power. Opinion must be manipulated continuously for the supporters to continue rooting for their team. Team Trump.

Just in case the seduced do not understand this compact, Trump tells them repeatedly that their loyalty is unshakeable; he could shoot someone on the streets of New York, he told them, and they would love him still. He can sexually assault women, and they love him still. He can abandon truth, morality, and obedience to the law, and still the spell is not broken. If he trips along the way, they will contrive not to notice, or contrive to see it as part of a clever plan—or perhaps a necessary compromise with dark forces. Those who have been drawn into the performance, who have identified with him, watch adoringly, with bated breath; the more vulnerable he is, the more he is their beloved, flawed Don Juan, or their death-defying trapeze artist, who will rise up again.

> I remember one day at the Opera [Clément (1989, p. 94) tells us]. The stage was immense and it was red; the characters in this sublime farce seemed miniscule and dwarfed by the space. Don Juan, gesturing grandly with his cape, caught his foot in a guitar, and came crashing down, ludicrously, head over heels, down the huge staircase on which he was delivering his famous serenade. Doubtless my memory amplifies the incredible noise. Doubtless the audience barely breathed a sound then, the way they do when an acrobat misses the trapeze; they concentrate on forgetting the accident, they blot it out.

Don Trump will make the country great by cleansing it; and with that he will return them to the American dream of riches and white male supremacy.

The dream doesn't have to be logical. If Baudrillard is right, it is more powerful in its non-sense. He can promote himself with false claims, and they will hold on to them as if they are true; he will deal with their anxieties, blowing away as fake news both global warming and the dire state of the economy. You have to believe in him, to believe in your own future.

As well as his admirers there are many willing to engage in critique. Trump frets about the negative press he receives from them (Haberman, Thrush & Baker, 2017). Only his representation as the biggest and the greatest and the most spectacular will do. It's an apparently hopeless desire for a man who attracts so much global derision. His wish for total adoration is perhaps a classic case of "cruel optimism" (Berlant, 2011, p. 1). He is both his own love object and the impediment to his universal adoration.

In stories of romantic love, we die for love—Lucia goes mad for love and kills her lover's usurper. But love can, also, be betrayed. Not only might the seducer grow bored with his audience, but loving audiences can turn away, bored. Guilt may ripple through their ranks, and suddenly "the imposing glamor of the man who stands up to God, to death, to women … is swallowed up in hell to ease the conscience of a jubilant body" (Clément, 1989, p. 94). The fall of Mugabe, brutal dictator for 35 years, gave rise in November 2017 to just such jubilation.

The Spectacle as Assemblage

It is tempting to individualize this spectacle that we wake up to every day; to see Trump as the originator or source of it all rather than analyze the multiplicities and mutual entanglements. It is easy enough to characterize him as a rogue individual, as Pussy-Grabber-in-Chief, or as Kim Jong Un names him, an old and frightened barking dog. The alternatives, that he is an expression of something deeply held and desired in the American psyche, or that he is a manifestation of the faltering and imminent collapse of capitalism, "a walking example of American decline" (McDonald, 2017, p. 12), as the Chinese characterize him, are more difficult to contemplate.

Whichever way he is characterized, it is much easier to think that he is the problem. It is also tempting to blame the people who voted for him, as if they too were not entangled, contradictory multiplicities.

Those who voted for him were not, of course, necessarily voting for Trump. While some were seduced by his projection of himself, others voted out of habit, for the side they always voted for, though Trump seemed at first a strange fit with Republican ideology, which had successfully disguised its commitment to white male supremacy and rapacious capitalism. Some were voting against Hillary, having swallowed the Russian propaganda along with Trump's amplification of it. They may also have been looking for a line of flight, a break with custom, or mud to throw in the face of the political class whose time was up: "Let's get someone more like us." Any old rich non-conformist Don Juan or Don Trump could arguably have filled that slot.

Although it is Trump who constantly courts attention, the spectacle is woven not just out of this one Don Juan and his followers. The mirrors he holds up to his voters, and the mirrors they hold up to him create a hall of mirrors, without any original. There are endless distorted repetitions, made up out of fragments of neoliberalism, rapacious capitalism, white male supremacy, the dying flickers of democracy, and the shards of deep-seated fears and concerns. This hall of mirrors has cracked and distorted glass, breaking the image up, so the whole is almost impossible to see. It takes some delving into history to get a sense of what kind of pattern the fragments might fall into.

Neoliberalism and Its Force in Shaping the Present

The Trilateral Commission was founded in 1973 by David Rockefeller, Chase Manhattan Chairman, Zbigniew Brzezinski, President Carter's national security advisor, and a cabal of other eminent power brokers. From their perspective, it was not foreign powers, or the internecine war between political factions, but the mobilized citizens who were the problem. Sklar quotes their report on the problem of our governability, published in 1975:

> *The vulnerability of a democratic government* in the United States (thus) comes not primarily from external threats, though such threats are real, nor from internal subversion from the left or the right, although both possibilities could exist, but rather from the internal dynamics of democracy itself in a highly educated, mobilized, and participant society. [Italics ed.]. *The Crisis of Democracy: Report on the Governability of Democracies to the Trilateral Commission.*
>
> *(Sklar, 1980, p. 3)*

The Trilateral governments had sustained numerous shocks from within: "militant protest and sustained political mobilization shook the stability" of governmental power (Sklar, 1980, p. 6). The people, that is, the people like us, they believed, had become ungovernable, causing a "decline in the strength of imperialism abroad" (Sklar, 1980, p. 43). In their report, called "The Crisis of Democracy," they wrote:

> The democratic surge threatens the fabric of international relations by undermining the U.S. role as global police. A "government which lacks authority and which is committed to substantial domestic programs will have little ability, short of cataclysmic crisis, to impose on its people the sacrifices which may be necessary to deal with foreign policy problems and defense" (such as the draft and social welfare cutbacks). Public awareness and opposition put a brake on U.S. intervention in the post-Vietnam era. Only strong government (i.e. *without strong opposition and oversight*) can effectively rule in the interest of international capitalism.
>
> *(Sklar, 1980, p. 43 quoting "The Crisis of Democracy" by Crozier et al., the Trilateral Task Force on the Governability of Democracies, emphasis added)*

Some 300 members of the Trilateral Commission, Sklar noted, included men drawn from international business and banking, from academia, the media, and conservative labor. Their common purpose was:

> to engineer an enduring partnership among the ruling classes of North America, Western Europe, and Japan—hence the term "trilateral"—in

order to safeguard the interests of Western capitalism in an explosive world ... *"trilateralism" refers to the doctrine of world order advanced by the Commission.*

(Sklar, 1980, p. 2 emphasis added)

The origin of neoliberal ideology is generally posited as Adam Smith. But there is something in the current spectacle that draws attention to darker, messier roots. What the fragments in the hall of mirrors begin to hint at is that the new world order being created globally by this international cabal of ruling class men, had historical roots in fascism and in white male supremacy, particularly as they had evolved leading up to and during WWII. The alliance with Nazism goes back at least as far as the aftermath of WWI. Henry Ford, for example, bought *The Dearborn Independent* newspaper in 1919 and filled it with anti-Jewish propaganda, further publishing that propaganda in a book called *The International Jew: The World's Foremost Problem* together with *The Protocols of the Elders of Zion*. One of the readers he inspired was to become the leader of Hitler's youth; another was Hitler himself.

> Although Ford made a public apology in 1927, admitting that *The Protocols of the Elders of Zion* was based on lies and forgery, in 1938 he received the highest Nazi order that could be conferred on a non-German, the Order of the German Eagle, together with a personal greeting from Adolf Hitler, who, incidentally, had a portrait of Henry on the wall of his Munich Office.
>
> *(Åsbrink, 2016, p. 57)*

The alliances crossed yet other surprising lines. Haj Amin al-Husseini, the Grand Mufti of Jerusalem, for example, had for some time before WWII been in cahoots with Hitler, who had made him an honorary German Aryan. They shared a fascination with the creation of outsiders, and the means of eliminating them. As Åsbrink says:

> Hitler's soldiers read [the Grand Mufti's] writings about the enemy within that had to be extirpated. He recruited at least 20,000 Bosnian Muslims to the SS. From 1941 to 1945, he lived in Berlin, met his friend Hitler, and discussed his vision of solving the Jewish problem in the Middle East in the same way as in Europe.
>
> *(Åsbrink, 2017, p. 79)*

In 1947 trials for murder of Jews, Roma, homosexuals, members of the Resistance movement, and mentally ill people committed by the Nazi regime were underway. But many Nazis had escaped to South America and Eastern Europe to revive white male supremacy when the time was ripe. Argentina's President

Peron, for example, recruited 1,000 Aryan Nazis. Lists were compiled of Danes, Norwegians and Swedes who had worked for the Nazis. They were sent passports and visas. Argentine diplomats in Stockholm and Copenhagen assisted in the stealing and forging of passports and in the concealing of Nazi identities (Åsbrink, 2017, p. 71). The British released 500 Nazi prisoners on the grounds that it was too expensive to keep them (Åsbrink, 2017, p. xx).

Nazism was not defeated in WWII; rather, it went underground and emerged elsewhere in new guises. Neoliberalism was one of those guises, though who it nominates as outsiders shifts with political expediency.

The task the Trilateral Commission set itself was to generate strategies suitable for the US and allied nations to further their project of the domination and control of their own people who had become ungovernable, with their demands for social welfare programs and progressive movements, including protest movements undermining US military imperialism. Those of us who were passionately fighting for social justice had become the enemy within.

The FBI and CIA counter-intelligence programs were used to "disrupt and destroy progressive movements in the 1960s and 70s; [including the use of] systematic police brutality against Chicano/as, Blacks, Native Americans and Puerto Ricans" (Sklar, 1980, p. 6).

In the 1970s, the Trilateral Commission looked for strategies that would do that same work without the citizens realizing what was being done to them. Through the clever insertion of new discourses and practices, workers would willingly exploit themselves as they shed the privileges and protections their predecessors had won for them over previous decades—predecessors like Martin Luther King.

Or predecessors like Eleanor Roosevelt, who met, post WWII, with her working group at Lake Success, to work on a counter-discourse that would put a stop to Nazi and other white male supremacist forms of violence. Together her group collated the last two centuries' ideas about human value and dignity, creating a 400-page document:

> [In 1947] Eighteen delegates from 16 countries are tasked with distilling a few drops of truth from the thousands of ideas extracted from earlier traditions. The word shining star-like over their joint and future efforts, leading them on, is universality.
>
> The Confucian philosopher Mencius is quoted for his 2000-year-old precept: people matter most. The state is of less consequence. The ruler is least important.
>
> Hindu thinkers and their high-flown thoughts are cited: freedom from violence, from greed, from exploitation, from humiliation, early death and sickness, the absence of intolerance, fear and despair.
>
> *(Åsbrink, 2017, pp. 95–96)*

What they produced was a philosophy of hope, not focused on violent cleansing and destructive forms of control, but on life, and on bringing quality of life to all humanity.

Another of our predecessors was Raphael Lemkin, who, in 1947, abandoned everything in his struggle to have genocide recognized as an international crime. He dedicated his life to the single resolve: "to constrain the evil of the world" (Åsbrink, 2017, pp. 106–107). For that he needed the UN, and so he left his job in Washington, gave up his substantial salary, and moved into a room in Manhattan. He was tireless in lobbying those who had influence. A decade later, in 1957, it was still possible to be convicted of a single murder, and to "remain unpunished for annihilating a whole group" (Åsbrink, 2017, pp. 106–107). Lemkin devoted his life to having genocide recognized as a crime against humanity, a crime for which one could be tried in the Hague.

The energy, the excitement, the passionate belief in the possibility of a better world created a force to be reckoned with. That reckoning came in the shape of the Trilateral Commission. And despite that reckoning, we are here still, though struggling under the effects of the neoliberalization of education and of the academy.

Piecemeal Implementation

The invention of neoliberal discourses and practices was a secretive affair, and its implementation deliberately piecemeal. One of its strategies, designed to fool those it sought to bring under control, was the use of our own language to suggest that what they were "really" doing was not bringing us under control, but giving us what we wanted—the means of demonstrating the quality of our work, and recognition of our ethical accountability. It took decades to work out where neoliberalism was coming from, and where it was headed. Reagan and Thatcher, and Howard in Australia, were among the first to buy into its premises. Guided by members of the Trilateral Commission, they set out to increase their controlling power and to contain the ungovernable. Citizens were to be reined in, and to be transformed into producers and consumers, no longer citizens, but a generic, replaceable, *homo economicus*. These ideas were not entirely new in 1960 and 1970, but the strategies for implementation were.

In order to survive in the new economic order, in which "society" was to be abandoned, each individual was made into a self-interested, entrepreneurial, narcissistic subject, who loves itself in excess, in much the same way misers love their money in excess (Lord, 2013). That love of self was to be generated by splitting individuals off from the social fabric, causing them to feel vulnerable in the face of ever more limited economic resources that they must compete for. The new neoliberal subject was to be weaned off social values, replacing those values with the single value of capital. Each was to be fed the fantasy (with a little hyperbole) of making excessive amounts of money while starting with nothing.

Eventually that generic neoliberal subject, in 2016, would come to desire an unprincipled, flamboyant money-maker as its king. Because neoliberal subjects, by design, were no longer ethical subjects who could be trusted to care about the quality of their work, state and institutional controls and surveillance were put in place. Professional workers, in particular, were no longer to be trusted, since they were, arguably, the most likely to engage in critique; they too must be put under surveillance and subjected to bureaucratic controls. Within each institution, reduced funding was used as a driver to dislodge individuals from social bonds and commitments. The plan was to weaken collegiality through heightening competition, through attacks on unions, through thorough restructuring and re-naming of groups, so that shared histories and the bonds of collegiality were lost.

The unprincipled subjects created through neoliberal strategies and structures no longer care about the question of who is controlling them, since they have been manipulated into defining themselves as free. Foyster, writing about the Facebook and Cambridge Analytica scandal observes:

> The cynical marketers are right. We always suspected Facebook was a voracious machine to monetise our personal data, but we uploaded ourselves anyway, trading privacy for little dopamine hits of peer affirmation. The real question isn't how this happened, it's why we knew but didn't care.
>
> *(Foyster, 2018, p. 3)*

The creation of unprincipled subjects, no longer engaging in critique and no longer committed to values other than economic values, ruled by other unprincipled subjects, creates precisely the conditions for Trump's rise to power. A further, deeply ironic effect of the Trilateral Commission's work was that the creative life of workers was damped down by these petty-bureaucratic controls, thus depriving capitalism of one of its most vital resources (Davies *et al.*, 2006a and b).

And so ...

There was a surge of energy, post WWII, that went into creating a more just world; we were both the beneficiaries and the participants in that surge; governments were to be confronted, and confront them we did. Qualitative research in the social sciences was a vital part of that movement—that social change. The power of qualitative research should not be underestimated. The Trilateral Commission saw the danger that we, along with critical journalists, posed for governments, and their strategies were aimed at us. Sklar wrote in 1980, about the governability report, and quoting from it:

Trilateralists point to the media as one opposition power threatening the governability of democracy. An adversary culture among intellectuals (and student followers) is believed to be another. So-called "value-oriented intellectuals" [the report said, were traitors] ... in action as well as in words, taking part in causes like the antiwar movement, and, currently, the antinuclear movement. "This development," states the governability report, "constitutes a challenge to democratic government which is, potentially at least as serious as those posed in the past by the aristocratic cliques, fascist movements, and communist parties." Upholding the values of the corporate system, in contrast, are the "technocratic policy-oriented intellectuals" who should be cloned on a global scale [as they subsequently were, through the Harvard MBA].

(Sklar, 1980, p. 40)

The Trilateral Commission saw our work as dangerous. No Child Left Behind tried to eradicate it. They would not have gone to so much trouble if our work was of no consequence.

Our task, then, is not one for the faint-hearted; I believe we need to develop much stronger political and historical strands in our work, so we are not naïve about our own power, or the forces that line up against it. Trump has done us a favor in drawing our attention to politics, and in making the workings of neo-liberalism and rampant capitalism more visible. Our work matters, perhaps more than we realize. We must offer clear thinking and informed critique, and a different model of humanity than the generic neoliberal subjects we have all been pressed, one way or another, into becoming. The movement we are currently engaged in, of opening up post-humanist and new materialist concepts, takes us toward greater response-ability to the non-human world, recognizing our shared material vitality with it; and toward a dismantling of the human hubris of narcissistic individualism and of anthropocentrism. These are vital shifts toward countering neoliberalism and its "legacies of white supremacy and rapacious capitalism" (Kim & Yancy, 2017, np).

References

Åsbrink, E. (2017). *1947. When Now Begins*. Melbourne: Scribe.

Baudrillard, J. (1990). *Seduction* (trans. M. Singer). New York: St Martin's Press.

Berlant, L. (2011). *Cruel Optimism*. Durham, NC: Duke University Press.

Clément, C. (1989). *Opera or the Undoing of Women* (trans. B. Wing). London: Virago.

Crozier, M., Huntington, S. P., & Watanuki, J. (1975). *The Crisis of Democracy*. New York: New York University Press.

Davies, B. (2014). *Listening to Children. Being and Becoming*. London: Routledge.

Davies, B., Browne J., Gannon, S., Honan, E., & Somerville, M. (2006a). Embodied women at work in neoliberal times and places. In Davies, B. and Gannon, S. (Eds.) *Doing Collective Biography*, Maidenhead: Open University Press. 61–78.

Davies, B., Browne J., Gannon, S., Honan, E., & Somerville, M. (2006b). Truly wild things: interruptions to the disciplinary regimes of neo-liberalism in (female) academic work. In Davies, B. and Gannon, S. (Eds.) *Doing Collective Biography*, Maidenhead: Open University Press. 79–87.

Faulkner, D. (2017). Gospel truths. *The Saturday Paper* November 18–24, p. 20.

Foyster, G. (2018). Facebook, unmasked. *The Saturday Paper* March 31–April 6, p. 3.

Haberman, M., Thrush, G., & Baker, P. (2017). Inside Trump's hour-by-hour battle for self-preservation. *New York Times* December 9.

Kim, D. K. & Yancy, G. (2017). An Open Letter of Love to Kim Jong-un, *The Stone*, November 13.

Lord, C. (2013). *Aristotle's Politics: Second Edition*. Chicago: University of Chicago Press.

McDonald, H. (2017). China ready to take advantage of Trump. *The Saturday Paper* November 4–10, p. 12.

McKenzie-Murray, M. (2017). Battle him of the republic. *The Saturday Paper* March 11–17, pp. 1, 10–11.

Massey, D. (2005). *For Space*. London: Sage.

Sklar, H. (1980). *Trilateralism. The Trilateral Commission and Elite Planning for World Management*. Boston: South End Press.

Trump, D. & Schwartz, T. (1987). *The Art of the Deal*. New York: Random House.

13

STITCHING TATTERED CLOTH

Reflections on Social Justice and Qualitative Inquiry in Troubled Times

Karen M. Staller

This address was delivered as one of two keynote performances opening the fourteenth Annual International Congress of Qualitative Inquiry (ICQI) in Champaign-Urbana, Illinois on May 17, 2018. The Congress, held in the Illini Union on the campus of the University of Illinois, sits on the land once occupied by a confederation of native tribes known collectively as the Illini.[1] The Congress always opens with recognition of this history and a welcoming ceremony performed by indigenous people.

ICQI has a devoted following. In attendance are between 1,200 and 1,400 delegates from sixty nations representing over twenty-five disciplines. Obviously, a Congress of this size and scope is the result of scores of influential people who have shaped the field of qualitative inquiry for decades as well as the dozens of up-and-coming scholars who have often served in administrative and organizing roles. Nonetheless, arguably the single most notable figure in bringing this entire vision to life since its inception is Norman K. Denzin, who has played an outsized role with respect to qualitative inquiry broadly and the ICQI more specifically.

This keynote address was written with its oral performance in mind. It was accompanied by visual images reinforcing its message. For future readers, perhaps the most significant contextual information I can provide is that Donald J. Trump was elected as the forty-fifth president of the United States on November 8, 2016. President Trump took office on January 20, 2017. My invitation to speak at the Congress arrived from Mr. Denzin on January 31, 2017. President Trump had been in office a mere eleven days. The theme announced for the 14th Congress was "Qualitative Inquiry in Troubled Times." Significantly, Mr. Denzin requested a keynote title and an abstract of its content by March 15, 2017, which was two months into the Trump Administration. The keynote address itself was scheduled for delivery a year and four months after President Trump took office. Given the context, committing to the content of an address to be delivered a year later seemed fraught. So in my abstract I offered to tailor my presentation from a diary of thoughts collected during the year. This is what happened in the process.

The keynote address follows.

Introduction

In 1880, social reformer Charles Loring Brace wrote,

> It is remarkable how little can be known in each generation of those [who came] before [and] who have done the most to make the present better than the past. The best workers for human good seem to be silent.[2]

It is clear that Donald J. Trump will never be remembered as one of those silent workers. He has spent much of the last year producing a Twitter storm of acrimonious and insolent tweets stirring up divisiveness and confusion in his first year of the US presidency.

For me, much of the past year has been devoted to two projects. One leading me here today in which I promised a talk about mending tattered cloth in troubled times. The second was the culmination of nearly a decade's worth of research on Mr. Brace and his work founding the Children's Aid Society of New York in 1853.[3]

However, the more time I spent collecting pieces for my tattered cloth project ... such as articles on:

- the plight of 65 million desperate refugee world-wide[4];
- the tragic Grenfell fire in London, a towering inferno that consumed a 24-story apartment building housing mostly poor families[5];
- the aftermath of Hurricane Maria which has left much of Puerto Rico without power, water, or basic necessities months after the storm[6];
- the riot in Charlottesville, Virginia, where white nationalist and neo-Nazis groups chanted racist and anti-Semitic slogans and rallied around Confederate monuments honoring the American Civil War[7]

... and the more I jotted down notes on Muslim bans and banned words or on President Trump's chaotic approach to governance, the more inspiration I drew from Mr. Brace's steady, courageous, and visionary leadership founding the Children's Aid Society. I began seeing similarities between his times and our own. In fact, the two projects began to morph into one. So today I will talk about tattered cloth. Not based on a single distracting year but across centuries.

I should start with three disclaimers, particularly in light of our beautiful welcoming ceremony from the Indigenous Inquiries Circle remembering those whose land we currently occupy. Mr. Brace was an American of European descent and his work is part of that slice of American history.

Second, I am not holding him out as a saint or as immune from biases of his day, in particular as they related to indigenous peoples. Nonetheless, as a radical progressive thinker in his own way, laying down a foundational base for what became the profession of social work, he cared deeply about the children of immigrants that others found vicious, dangerous, and threatening.

My final disclaimer is that my profession of social work makes no apologies for taking value-based positions.[8] Its central mission is one of promoting social justice. That doesn't mean we have always gotten things right. In fact, we often haven't. Nonetheless, in our failures we also learned.

So, let me tell you a little bit about Mr. Brace and why I do think his work is significant for this Congress and beyond.

Mr. Brace's Background

For some his name may be familiar. In modern scholarship Mr. Brace's work has been credited, and criticized, as being a precursor to our child welfare or foster care system. Some have dubbed his efforts the "orphan trains" because of the practice of removing poor, mostly immigrant children from cities like New York and "placing them out" in families in the western United States.[9] I will argue this characterization of Mr. Brace's work does him a disservice for being too narrow. His vision was far larger and his impact far greater.

Charles Loring Brace was born in 1826 and died in 1890. His life spanned eras in American history we now call the antebellum period, the American Civil War, Reconstruction, and the Gilded Age. Let me tackle six consequences in reverse chronological order.

First, toward the end of the nineteenth century, the level of income inequality was unprecedented. Rapid industrialization, among other things, produced a spectacularly wealthy class known as the "robber barons."[10] On the other side was extreme and wrenching poverty.

Second, the numbers immigrating to the US were also unprecedented.[11] Europeans were fleeing famine, wars, political, and religious oppression. It is worth highlighting that between 1845 and 1852 the Irish potato crop failed continuously for seven years. "A million people died of starvation" and hundreds of thousands fled.[12] For those already settled in the US the influx of desperate Irish immigrants, in particular, seemed ominous.

Third, as if the ethnic difference wasn't intimidating enough, large numbers of immigrants—particularly the Irish and Italians—were Catholic. Many here feared that Catholicism was autocratic rather than democratic in nature because of its leadership by a Pope in Rome. They argued that "Protestantism defined

American society," and that Catholicism was not compatible with basic Americans values.[13]

Fourth, these fears gave rise to the Know-Nothing movement and a short-lived political party, called the "American Party."[14] These "nativists" (and I recognize the irony in the term) called for a twenty-one-year naturalization period for immigrants, Bible reading in public schools, deportation of foreign paupers and criminals, and elimination of Catholics from government office.[15]

Fifth, let me add to the mix the publication of a highly controversial book in 1860. Charles Darwin's *On the Origin of Species* set off a contentious debate that pitted science against religion.[16] Where did human beings come from? The book offered evidence, a grand theory, and new ways to understand the world.

Finally, and significantly, in US history was the issue of slavery. As the United States inched toward a Civil War which would pull the young Republic apart at its seams, it is important to remember that slavery was not universally understood as a moral issue. Many saw it through an economic or political lens. The Southern economy was completely dependent on slave labor but so too was Northern industry and shipping.[17] Politically, the same argument that had divided the colonies since before the American Revolution, a question of popular sovereignty, or states' rights, was at the fore. With the addition of each new state to the union, the delicately arranged balance between free and slave states was threatened.

So, let me summarize. Charles Loring Brace lived during troubled times. Issues at play included ethnic and religious tensions, mass migration and fear of immigrants, racial injustice, political division, acute income inequality and debates over the role of science and religion.

In his nineteenth-century world, I see direct parallels to our own troubled times. We are in a period of unprecedented global migration. We face newly invigorated nativist movements. Ethnic, religious and racial tensions have flared. Science, religion, morality and politics are linked. Income inequality is at record levels.[18] In fact, one in five children in the US lives below the poverty line.[19] Worse yet, of those, nearly half live in "extreme poverty," defined as living on less than $2.00 a day.[20] Or less than what many of us spent this morning on a single cup of coffee.

In short, we might ask, what could one person possibly do against these odds?

Mr. Brace: A Dangerous Man of Reason

Again, I turn to Mr. Brace but it is helpful to know something about the man himself. He came from a family of advocates and intellectuals.[21] His family tree included religious, political, judicial and scientific leaders, educators and scholars. For example, his brother-in-law was Asa Gray, a prominent botanist and early defender of Charles Darwin's work[22] and the Beechers, including

Charles Beecher[23] and Harriet Beecher Stowe, who wrote *Uncle Tom's Cabin*.[24] Mr. Brace drew from this entire mix when forging his own intellectual, religious, scientific and political beliefs.

This produced a fiercely independent free thinker in Mr. Brace. He abhorred dogma. He wrote that it was complicated but important to break free of conventional thinking or "what-we-always-have-been-educated-to-believe."[25] What others accepted on blind faith, Mr. Brace questioned.

Given his maverick and catholic approach to knowledge, it's not surprising his ideas were frequently characterized by others as being "dangerous." Yet in response, he argued the only truly dangerous thing was holding any belief too sacred or too cherished to be investigated. He resolved "never for a moment to refuse hearing a truth because it is new, and never to be afraid to dig under a belief because it is old and dearly loved."[26] He wrote,

> We may reason wrong; we may be prejudiced or foolish or weak, but that there can be anything wrong in searching for truth freely, or in uprooting the dearest opinion to see what lies under it, or in applying our individual judgment to any truth (be it even God's existence), I do not see.[27]

Mr. Brace relished debate and routinely sought out those with different ideas in order to test his own logic. His verbal jousting on politics, science, philosophy, religion, or ethics, could last for hours, exasperating his friends who complained he was "always disposed to pursue a debate through[out] the night."[28]

Finally, and significantly in his historical context, Mr. Brace was a radical abolitionist. For him the question of slavery was unequivocally a moral one. He viewed the African slave trade as "the most dreadful curse which has perhaps ever afflicted humanity."[29] In particular, he once wrote, "I do not wish to rant, but it is the deepest feeling of my heart, that no darker stain rests on this country than this slavery. Men *must* see it sometime"[30] (emphasis in the original).

In short, Charles Loring Brace operated according to a strong moral compass. He lived life by asking big questions. He drew his answers from eclectic sources and subjected them to rigorous testing.

Mr. Brace's New York

When he first arrived in New York in 1848 he faced some new social challenges created by the rapid urbanization. At the time, New York consisted of what the indigenous Lenape people called Manna Hata, or island of many hills. In the absence of bridges or tunnels connecting it, the island of Manhattan was trying to absorb the influx of poor immigrants in a constrained geographic space.

To do so, New York experimented with new residential housing for the poor called tenements. These unregulated, multi-family dwellings were exclusively designed to pack as many people into as small a space as possible. I can't over emphasize how deplorable these building were. Early inspectors found conditions "far exceeding the limits of previously conceived ideas of human degradation and suffering."[31] With no indoor plumbing, hundreds of people shared a single outdoor water pump and overflowing outhouses. With no regulations requiring windows, only coal heat and no electricity or gas, not only were living conditions overcrowded but sooty, dark, damp, and airless.

Given these conditions, it's not surprising that the mortality rates among the poor were extremely high.[32] Air- and water-borne diseases were common. As were industrial accidents and deadly fires. As a consequence, orphans were numerous. In addition, overcrowding also left poor children susceptible to eviction. The net result was rampant child vagrancy and homelessness. In addition, child labor was the norm.[33] Children worked in street trades, factories, and sweatshops. Others begged, scavenged, or stole to survive.[34] There was little time for formal schooling and their long-term prospects for getting out of poverty were relatively bleak.

In the public mind immigration, child vagrancy, poverty, petty crime, and immorality all seemed linked. Wealthy New Yorkers feared the "dangerous classes."

So What Was Mr. Brace's Plan?

So, what did Mr. Brace do? When he arrived in 1848 as a twenty-two-year-old he was immediately concerned about the conditions of the poor. Rather than being immobilized by the size of the problems he saw only a gigantic puzzle that needed solving. So he set to work.

Surveying the morass of issues before him, he quickly settled on *child poverty* as being the keystone of the whole. Mr. Brace was critical of the existing methods of dealing with the poor. He believed missionary work or what he called "the old technical methods—[of] holding prayer-meetings, and scattering Bibles," was dangerously outdated.[35]

In addition, the practice of locking up vagrant children in juvenile reformatories only taught what he called "technical virtues."[36] Even worse, the practice of disposing of poor children through indentured servitude contracts impinged on their free will and was too close to enslavement. In short, Mr. Brace believed entirely new methods were called for that were rooted in the moral values of religion but not constrained by existing social, political, scientific or religious conventions.

So in 1853 he joined a small group of like-minded individuals and launched the Children's Aid Society. As its first Secretary Mr. Brace had the job of giving it shape. His first order of business was to collect evidence. He visited every poor neighborhood in New York.[37] He talked directly to residents, parents, street children, clergy, police officers and employers. He observed conditions both during the day and in the middle of night. Then he meticulous compiled his results, reporting his findings to the public through written reports, circulars, newspaper articles, and Sunday sermons.

At the same time, he began experimenting with a variety of programs that would comprise the Children's Aid Society. It is worth noting that some of his earliest efforts were nothing short of a disaster.[38] Undeterred, he merely considered the lessons learned and tried something different. In short order he designed a grand plan, a blueprint for action. It rested on a single guiding principle. "The great idea of the Children's Aid Society," he wrote, [is] "to help the children help themselves."[39] Every aspect of its work must serve that core mission. In addition, he believed:

- Poor children needed the same kind of individualized attention as wealthy ones.
- Family placements were preferable to institutions.
- Education—broadly defined—was the most powerful tool in preventing poverty and crime.
- Meeting children's basic needs including food, clothing, shelter and sanitary conditions had to go hand-in-hand with any other intervention.
- Finally, he believed the country was preferable for children because:
  - rural living conditions were healthier than urban;
  - there was a demand for child labor that was oversupplied in the city;
  - there were greater life-long opportunities in the country, where poor children could enter what he called the "ruling class"; and
  - families in the west were willing to provide children homes, education and employment.

Based on these principles, Mr. Brace declared his plan "very simple." It was to operate "a Central Office, Agents to find the poor children, Schools to educate them ... Lodging-Houses to shelter, train, and clothe them, and Western Agents to convey them to homes in the West."[40] As simple as this may sound, it was actually stunning in its size, scope and vision.

The plan consisted of a seamless set of interventions that modern social workers would call community-based outreach, common education, vocational training, case management, job placements and referrals, short-term crisis shelter, transitional living facilities, foster care and adoption.

In the process, Mr. Brace helped build an entirely new profession of applied philanthropy, which blurred the boundaries between charity work and missionary work, as well as between religion, science, theory and practice.

Hurdles, Challenges and Practices

In spite of this grand vision, it was not smooth sailing. As the US entered a civil war, Mr. Brace was threatened with dismissal from the agency he had built from scratch. The Trustees worried his extreme abolitionist views would jeopardize the reputation of the Children's Aid Society.[41] In the end, he was permitted to stay but not without a fight.

In addition, critics repeatedly attacked the Children's Aid Society's approaches to working with poor children, saying it was "scattering the seeds of vice and corruption."[42] Significantly, Mr. Brace faced each wave of criticism with the same formula. He demanded to know what evidence his critics were using. He conducted his own independent investigations. And he responded with thoughtful, comprehensive and well-reasoned public reports.

Mr. Brace's Impact

So what did he accomplish? I'll turn to his last annual report of 1889.[43] That year, the Children's Aid Society placed out 3,000 boys and girls, but also men and women, in homes or employment situations in the western United States.[44] However, in New York City, it had also worked with an additional 38,853 children. It ran six lodging houses for homeless boys and girls, sheltering over 12,153.

It ran 21 community-based day schools and 12 night schools located in poor neighborhoods, educating 11,331 children. The schools included two ethnically-based programs, one for German children and another for Italians. The Society employed hundreds of staff, including 151 salaried teachers. It also ran health- and disability-related facilities. These included programs for "crippled" boys and girls, and health homes for mothers and children.

At every single facility, the Children's Aid Society provided, food, clothing, medical care and sanitary conditions including bathing facilities, and well-lit and ventilated rooms.

However, here I wish to make a bigger point that goes beyond the quantitative impact of a single year. Mr. Brace had been engaged in these efforts for nearly forty consecutive years. Arguably he had a profound longitudinal impact on the health and welfare of hundreds of thousands of children.

Mr. Brace's goals amounted to nothing short of eradicating poverty and homelessness, decreasing crime and delinquency, reducing illiteracy, reducing unemployment and improving child and maternal health outcomes. In the end, his aim was nothing short of changing the entire social fabric. When asked about the Society's success, Mr. Brace attributed it to two factors. The first was the quality of its dedicated workers and the second was the soundness of the principles upon which it was founded.[45]

The Congress

At this point you may be thinking, that's all well and good but so what? What does that have to do with our own troubled times, or the Trump administration, or this Congress, or qualitative inquiry or the politics of evidence? I say, quite a bit.

Try substituting the words "qualitative inquiry" for "child poverty."

Then, look around.

For fourteen years a band of like-minded scholars from around the globe have gathered here in the heartland of the rural United States. You have answered a value-based call with social justice and social action at its core. No idea has been too dangerous or too radical to try. Together, you have brilliantly and creatively blended science, humanities and the arts. You have resisted forces that would silence voices, or methods, or speech, or cultural practices. You have insisted on confronting the politics of evidence and resisted the dominance of oppressive power structures in the academy and beyond.

We do live in troubled times, and we face complex social problems. However, each of you has experimented with diverse and innovative methods and believed in the power of qualitative inquiry in generating solutions. In short, all the features I culled from Mr. Brace's life and work from nearly two centuries ago can be found right here in this room.

After all, what was it that Mr. Brace and his Society really did?

- Facing complex social problems, fearlessly, Brace identified a core agenda.
- He then built a Society of like-minded people around it.
- The entire project was based on a commitment to a moral vision with action at its core and dealt courageously with the politics of the times.
- It rejected the traditional approaches imposed by others and instead designed entirely new ones.
- The design was rooted in qualitative forms of inquiry such as interviews, ethnography, and participatory action, but it employed evidence culled from religion, philosophy, science, observation, experience, experimentation and performance (or practice).
- The Children's Aid Society disseminated its findings through public scholarship.
- It was a project and approach that met with plenty of resistance but fought back.
- In the process, Brace built an entirely new institution. Not only did it have a profound impact on child well-being. It employed entirely new methods of intervention and it envisioned the transformation of the social and political order.

I used the Children's Aid Society to illustrate the possibility of having an equally profound impact. I argue that we need to look beyond the momentary distractions of a chaotic year, and rather see the entirety of this project and envision its collective force and its long-term impact. And, of course, to find the grand architect and visionary thinker for *this* project you need look no further than this dais: Norman Denzin has been committed to this moral, ethical and political vision of qualitative inquiry as a transformative force for social justice work for a lifetime. He has been undeterred by critics and steadfast in his values and has pieced together a tapestry of talent. He has created a home for maverick thinkers, activists and scholars.

The bigger question for the rest of us is what to do with this gift? While we can use this Congress as a welcome retreat, a place to replenish our souls and find inspiration, remember it is a *congress*, and not a *conference*. It envisions a larger political agenda. Just read Norman Denzin's "Qualitative Manifesto: A call to arms."[46] The real challenge is in *harnessing* its power beyond this place and space. For carrying the mission forward during the upcoming year, I borrow from youth movements around the globe and—dare I say from Donald J. Trump himself—I argue for micro blogging up our own Twitter storm united under an ICQI banner with an eye toward capturing our collective unity and yoking the real power in this room.

Hashtag #ICQI plus whatever: New qualitative research, global action, tenure and promotion, politics of evidence, ethical dilemmas, indigenous circles, or autoethnography ... whatever! Consider each tweet as a stitch for mending tattered cloth. The fabric pieces rest with each and every one of you. As Dorothy from *The Wizard of Oz* discovered, there is no place like home. All you have to do is come "to know *this* place for the first time."[47]

In closing, I'll note that a colleague once said of Charles Loring Brace that his work was "an example of what may be done by large brains, a big heart, and rare common-sense concentrated on one worthy object."[48] The same might be said for Mr. Denzin and for this Congress. Enjoy its warm embrace for the next two days but remember your responsibility to it as well. It is our job (hashtag) to carry it forward. # ICQIcarryitforward

Notes

1 *Illinois Tribes*. "500 Nations, Tribes & People." www.500nations.com/.

2 Brace, Charles Loring (1880). "The silent workers." Addendum (pp. 449–457) *The dangerous classes of New York: And twenty years work among them*. 3rd Edition. New York: Wynkoop & Hallenbeck, p. 449.

3 For this work see Staller, Karen M. (forthcoming). *New York's newsboys: Charles Loring Brace and the founding of the Children's Aid Society*. Oxford: Oxford University Press.

4 United Nations High Commission for Refugees (UNHCR): United Nations Refugee Agency. Figures at a glance. Retrieved from: www.unhcr.org/en-us/figures-at-a-glance.html.

5 "Grenfell Tower: Firefighters search overnight with toll expected to rise." *Guardian*. (June 14, 2017). Retrieved from www.theguardian.com/uk-news/2017/jun/14/fire-24-storey-grenfell-tower-block-white-city-latimer-road-london. Kirkpatrick, David

D., Hakim, Danny and Glanz, James, "Why Grenfell Tower burned: Regulators put cost before safety." June 24, 2017. *New York Times*.

6 "Hurricane Maria updates: In Puerto Rico, the storm 'destroyed us'" in the *New York Times*, September 21, 2017. Retrieved from: www.nytimes.com/2017/09/21/us/hurricane-maria-puerto-rico.html; Giusti, Carlos and Weissenstein, Michael, AP "Corps leaving Puerto Rico with hurricane recovery unfinished." ABC News. May 18, 2018. Retrieved from: https://abcnews.go.com/International/wireStory/corps-leaving-puerto-rico-hurricane-recovery-unfinished-55252133.

7 Hanna, Jason, Hartung, Kaylee, Sayers, Devon M. and Almasy, Steve, "Virginia governor to white nationalists 'Go home … shame on you.'" CNN, August 13, 2017. Retrieved from: www.cnn.com/2017/08/12/us/charlottesville-white-nationalists-rally/index.html.

8 For example, in the National Association of Social Workers (NASW) Code of Ethics in the United States, one of the six enumerated values listed is "social justice," and the underlying ethical principle is that "social workers challenge social injustice." This includes pursuing "social change, particularly with and on behalf of vulnerable and oppressed individuals and groups of people." NASW Code of Ethics, Retrieved from: www.socialworkers.org/About/Ethics/Code-of-Ethics/Code-of-Ethics-English.

9 O'Connor, Stephen (2001). *Orphan trains: The story of Charles Loring Brace and the children he saved and failed*. New York: Houghton Mifflin. Holt, Marilyn Irvin (1992). *The orphan trains: Placing out in America*. Lincoln: University of Nebraska Press.

10 Beckert, Sven (2001). *The monied metropolis: New York City and the consolidation of the American bourgeoisie, 1850–1896*. Cambridge: Cambridge University Press.

11 In just 30 years, the total population ·of Manhattan rose from 197,112 in 1830 to 805,358 in 1860. (From Manual for the Use of the Legislature of the State of New York for the year 1857. Albany, NY: Weed, Parsons and Co., p. 201 and Report of the Council of Hygiene and Public Health of the Citizens' Association of New York upon the Sanitary Conditions of the City (1866). New York: D. Appleton, p. cxxi.)

12 Sowell, Thomas (1981). *Ethnic America: A history*. New York: Basic Books, p. 21.

13 Anbinder, Tyler G. (1992). *Nativism and slavery: The Northern know nothings and the politics of the 1850s*. New York: Oxford University Press, pp. 105–106.

14 Ibid.; Gorn, Elliott J. (1987). "Good-bye boys, I die a true American: Homicide, nativism, and working-class culture in Antebellum New York City." *Journal of American History*, 74(2): 388–410.

15 Ibid.

16 Fuller, Randall (2017). *The book that changed America: How Darwin's theory of evolution ignited a nation*. New York: Viking.

17 In addition to being a major shipping port, New York City was an industrial center with thriving clothing and cigar industries which were reliant on southern cotton and tobacco.

18 Saez, Emmanuel and Zucman, Gabriel (May 2016). "Wealth inequality in the United States since 1913: Evidence from capitalized income tax data." *Quarterly Journal of Economics*, 131(2): 519–578.

19 Child Trends Data Bank. "Children in poverty: Indicators of child and youth well-being." (Updated December 2016).

20 Edin, Kathryn J. and Shaefer, H. Luke (2016). *$2.00 a day: Living on almost nothing in America*. First Mariner Books; "Child Poverty in the United States." The Globalist, December 26, 2017. Retrieved from: www.theglobalist.com/united-states-child-poverty-social-welfare/.

21 Brace, Emma, ed. (1894/1976). *The life of Charles Loring Brace*. Reprint, New York: Charles Scribner's Sons.

22 Gray, Asa (undated). *Darwiniana: Essays and reviews pertaining to Darwinism*. Champaign, IL: Book Jungle; Dupree, A. Hunter, (1959) Asa Gray: American botanist, friend of Darwin. Baltimore: Johns Hopkins University Press.

23 Charles Beecher was a radical abolitionist. See Beecher, Charles (1851). *The duty of disobedience to wicked laws: A sermon on the Fugitive Slave Law*. New York: John A. Gray.

24 Reynolds, David S. (2011). *Mightier than the sword:* Uncle Tom's Cabin *and the battle for America*. New York: W.W. Norton.

25 Letter from C. L. Brace to Fred Kingsbury (November 1849). In Brace, Emma, ed., *The Life of Charles Loring Brace*. 1894. Reprint, New York: Charles Scribner's Sons, 1976, p. 78.

26 Brace, C. L. letter circa 1847, ibid., p. 39.

27 Brace, C. L. letter circa 1847, ibid., p. 38.

28 This complaint was lodged against Mr. Brace by Frederick Law Olmsted, a lifelong friend from childhood. Mr. Olmsted became known as the father of modern landscape architecture. Among his best-known works is Central Park in New York City. David Schuyler and Gregory Kaliss, eds., (2015), *The Papers of Frederick Law Olmsted: Volume IX: The Last Great Projects, 1890–1895*, Baltimore: Johns Hopkins University Press, p. 467.

29 Charles Loring Brace, (1882), *Gesta Christi: or A History of Humane Progress Under Christianity*, 4th edn, New York: A.C. Armstrong & Son, p. 364.

30 Letter from C. L. Brace to his father, John Pierce Brace, (April 11, 1849) in Brace, Emma, ed., *The Life of Charles Loring Brace*, p. 72.

31 Report of the Select Committee Appointed to Examine the Conditions of Tenant Houses in New York and Brooklyn. Report transmitted to the Legislature, Albany, NY: C. Van Benthuysen, Printer to the Legislature, March 9, 1857, p. 3. See also Report of the Council of Hygiene and Public Health of the Citizens' Association of New York upon the Sanitary Conditions of the City. New York: D. Appleton and Company, 1866. Anbinder, Tyler (2001), *Five Points: The 19th-Century New York City neighborhood that invented tap dance, stole elections and became the world's most notorious slum*. NY: A Plume Book. Dolkart, Andrew S., (2012), *Biography of a Tenement House in New York City*, 2nd edn. Chicago: The Center for American Places at Columbia College Chicago.

32 For example, in one "fever nest"—an area designated from the high rates of contagious diseases—one doctor reported 42 individuals died in five closely packed tenement building in a three week period. None of those who had taken sick recovered. Report of the Council of Hygiene and Public Health of the Citizens' Association of New York upon the Sanitary Conditions of the City. New York: D. Appleton and Company, 1866, 175–177.

33 Jeremy P. Felt, (1965), *Hostage of Fortune: Child Labor Reform in New York State*, Syracuse, NY: Syracuse University Press.

34 Matsell, George Washington (1849). "Embryo Courtezans and Felons": New York Police Chief George W. Matsell Describes the City's Vagrant and Delinquent Children, Reprinted in Harris, Thomas L. (1850, January 13). *Juvenile Depravity and Crime in our City: A Sermon by Thomas L. Harris*, New York: Charles B. Norton.

35 Charles Loring Brace, (1880), *The Dangerous Classes of New York and Twenty Years' Work Among Them*, 3rd edn, New York: Wynkoop & Hallenbeck, p. 76.

36 Charles Loring Brace, (1859), *The Best Method of Disposing of Our Pauper and Vagrant Children*, New York: Wynkoop, Hallenbeck & Thomas, p. 5.

37 Charles Loring Brace, *Early Diary of Charles Loring Brace, Founder of the Children's Aid Society* (unpublished, February 10, 1853–September 1855). *Records of the Children's Aid Society*, Series IV, vol. 38, New York Historical Society.

38 Among Mr. Brace's earliest doomed experiments was establishing a shoe pegging workshop designed to train boys in the craft. He quickly learned that training children in specific industrial activities was not helpful in the long run. Unlike the European counterparts, which relied on industrial schools for poor children focused on specific work activities, Mr. Brace believed the US had such a widespread shortage of labor in all areas that children merely needed a common school education and some basic skills to thrive in the long run.

39 Thirty-Seventh Annual Report of the Children's Aid Society, (1889), New York: Wynkoop, Hallenbeck & Co., p. 4.

40 Ibid, p. 5.

41 Brace, C. L. Letter "To a Trustee," (August 7, 1864) in Brace, Emma, ed. *The Life of Charles Loring Brace*, pp. 263–64.

42 Sixth Annual Report of the Children's Aid Society, (1859), New York: Wynkoop, Hallenbeck & Thomas, p. 37; Minutes of the Board of Trustees, November 2, 1859 and December 21, 1859, vol. 2, 1853–1861, in Records of the Children's Aid Society 1836–2006, New York Historical Society Archives; Brace, Charles Loring, (1880). *The Dangerous Classes of New York: And Twenty Years Work among Them*, 3rd edn, NY: Wynkoop & Hallenbeck, p. 235.

43 Thirty-Seventh Annual Report of the Children's Aid Society, (1889), New York: Wynkoop, Hallenbeck & Co.

44 CAS placed out 2,210 boys, 977 girls, 132 men, and 232 women in 1889.

45 Thirty-Seventh Annual Report of the Children's Aid Society, (1889), New York: Wynkoop, Hallenbeck & Co., pp. 3–4.

46 Denzin, Norman K. (2010). *Qualitative Manifesto: A Call to Arms*. London and New York: Routledge.

47 Paraphrasing the common refrain used by Norman Denzin to open the Congress each year taken from T.S. Eliot's "Four Quartets" (1942):

> We shall not cease from exploration
> And the end of all our exploring
> Will be to arrive where we started
> And know the place for the first time.

48 William E. Dodge, Letter (December 9, 1890) in Brace, Emma, ed. *The Life of Charles Loring Brace*, p. 484.

CONTRIBUTORS

Editors

Norman K. Denzin is Distinguished Emeritus Professor of Communications, College of Communications Scholar, and Research Professor of Communications, Sociology, and Humanities at the University of Illinois at Urbana-Champaign, USA. One of the world's foremost authorities on qualitative research and cultural criticism, he is the author or editor of more than two dozen books, including *The Qualitative Manifesto; Qualitative Inquiry Under Fire; Reading Race; Interpretive Ethnography; The Cinematic Society; The Voyeur's Gaze; The Alcoholic Self;* and a trilogy on the American West. He is past editor of *The Sociological Quarterly,* co-editor (with Yvonna S. Lincoln) of five editions of the landmark *Handbook of Qualitative Research,* co-editor (with Michael D. Giardina) of 15 books on qualitative inquiry, co-editor (with Lincoln) of the methods journal *Qualitative Inquiry,* founding editor of *Cultural Studies ↔ Critical Methodologies* and *International Review of Qualitative Research,* editor of four book series, and founding director of the International Congress of Qualitative Inquiry.

Michael D. Giardina is Professor of Media, Politics, and Physical Culture and Associate Chair in the Department of Sport Management at Florida State University, USA. He is the author or editor of more than 20 books, including *Sport, Spectacle, and NASCAR Nation: Consumption and the Cultural Politics of Neoliberalism* (Palgrave, 2011; with Joshua Newman), *Qualitative Inquiry—Past, Present, & Future: A Critical Reader* (Routledge, 2015; with Norman K. Denzin), and *Physical Culture, Ethnography, & The Body: Theory, Method, & Praxis* (Routledge, 2017; with Michele K. Donnelly). He is Editor of the *Sociology of Sport Journal,* Special Issues Editor of *Cultural Studies ↔ Critical Methodologies,* editor of three book series with Routledge, and the assistant director of the International Congress of Qualitative Inquiry.

Authors

Kakali Bhattacharya is Professor in the Department of Educational Leadership at Kansas State University. She is the author of *Fundamentals of Qualitative Research: A practical guide* (Routledge, 2017) and *Power, Race, and Higher Education: A cross-cultural parallel narrative* (Sense, 2016; with Norman Gillen).

Liora Bresler is Professor of Curriculum & Instruction in the College of Education at the University of Illinois, Urbana-Champaign. She is the author or editor of numerous books, including *Research in International Education: Experience, theory, practice* (Peter Lang, 2022), *International Handbook of Research in Arts Education* (Springer, 2007), and *International Handbook of Creative Learning* (Routledge, 2011). The impact of her work has resulted in a number of honors including a National Art Education Association Distinguished Fellow (2010), Outstanding Recognition Award by the American Alliance for Theatre and Education (2007), and the Ziegfeld Award for distinguished international leadership in art education by USSEA (2007). Her work has been translated into German, French, Spanish, Portuguese, Slovenian, Lithuanian, Hebrew, Chinese, and Korean, and she has given 40+ keynotes and 100+ invited talks in 30+ countries in North and South America, Europe, Asia, Australia and Africa.

Svend Brinkmann is Professor of Psychology and co-director of the Center for Qualitative Studies at Aalborg University, Denmark. He is the author or editor of more than 20 books, including *InterViews: Learning the craft of qualitative research interviewing* (with Steiner Kvale; Sage, 2014), *Qualitative Inquiry in Everyday Life* (Sage, 2012), and *Qualitative Interviewing* (Oxford University Press, 2013).

Julianne Cheek is Professor of Nursing at Østfold University College, Norway, and the University of South Australia. One of the leading qualitative health researchers in the world, she is the author of *Postmodern and Poststructural Approaches to Nursing Research* (Sage, 2000). She also holds a number of honorary professorships in South Africa and the United Kingdom, and serves on the editorial board of academic journals including *Global Qualitative Nursing Research*, *Qualitative Health Research*, and *International Review for Qualitative Research*. She is currently completing a book on the political economy of qualitative inquiry in the research marketplace.

Bronwyn Davies is a professorial fellow at Melbourne University, Australia, and an independent scholar. She is the author of numerous books, including: *Listening to Children: Being and Becoming* (Routledge, 2014); *Frogs and Snails and Feminist Tails: Preschool Children and Gender* (Hampton, 2003); and *Judith Butler in Conversation: Analyzing the Texts and Talk of Everyday Life* (Routledge, 2007).

Jennifer Esposito is Professor in the Department of Educational Policy Studies and an affiliate faculty member in the Alonzo A. Crim Center for Urban Educational Excellence and the Institute for Women's, Gender and Sexuality Studies at Georgia State University. Her research interests include urban education; race, class, gender and sexual orientation identity construction; qualitative methodology; and popular culture as a site of education. Her research has been published in several books and scholarly journals, including *Qualitative Inquiry, Equity and Excellence in Education* and *Educational Studies*.

Venus E. Evans-Winters is Associate Professor of Education at Illinois State University. She is the author of numerous books, such as *Teaching Black Girls: Resiliency in urban classrooms* (Peter Lang, 2005), *Black Feminism in Education: Black women speak back, up, and out* (with Bettina Love: Peter Lang, 2014), and *Re(teaching) Trayvon: Education for racial justice and human freedom* (Sense, 2014).

Aitor Gómez González is Associate Professor of Research Methods at the Univeristy Rovira I Virgili (Tarragona, Spain). He has published in the main journals on methodology, such as *Qualitative Inquiry* and the *Journal of Mixed Methods Research*, among others. He has been the main director of the PER-ARES Project (FP7, 2010–2014) at the URV aimed at strengthening the connection between SSH and society. He is currently the principal investigator (PI) of "SALEACOM: Overcoming Inequalities in Schools and Learning Communities: Innovative Education for a New Century," a Marie Skłodowska-Curie Research and Innovation Staff Exchange (RISE) with Stanford University aimed at extending successful educational actions across educational systems.

Anne Harris is currently Vice Chancellor's Research Fellow and Australian Research Council Future Fellow in the School of Education at Royal Melbourne Institute of Technology, Australia. She is a native New Yorker and has worked professionally as a playwright, teaching artist and journalist in the USA and Australia. She holds both master's and bachelor's degrees from New York University where she studied playwriting and performance with Tony Kushner, Arthur Miller, Peggy Phelan and Eve Ensler among others. In addition to her plays, she has published over 50 articles and six books that address the arts, culture, and performance, including her latest book *Critical Plays: Embodied Research for Social Change* (Sense, 2014).

Stacy Holman Jones is Professor in the Centre for Theatre and Performance at Monash University, Australia. She is the author/editor of eight books: *Kaleidoscope Notes: Writing Women's Music and Organizational Culture* (AltaMira, 1998), *Torch Singing: Performing Resistance and Desire from Edith Piaf to Billie Holiday* (AltaMira, 2007), *Handbook of Autoethnography* (Left Coast Press, 2013 co-edited with Tony E. Adams and Carolyn Ellis) *Autoethnography* (Oxford

University Press, 2015, co-authored with Tony E. Adams and Carolyn Ellis), *Stories of Home: Identity, Place, Exile* (Lexington, 2015, co-edited with Devika Chawla), *Writing for Performance* (Sense, 2016, co-authored with Anne M. Harris) and *The Handbook of Performance Studies* (Wiley Blackwell, forthcoming). She is the recipient of several research and teaching/mentoring awards, including the Janice Hocker Rushing Early Career Research Award and the Organization for the Study of Communication, Language, and Gender Feminist Teacher/Mentor Award.

Alecia Youngblood Jackson is Associate Professor of Educational Research in the Department of Leadership & Educational Studies at Appalachian State University. She is the author of *Thinking with Theory in Qualitative Research: Viewing Data Across Multiple Perspectives* (with Lisa A. Mazzei; Routledge, 2012) and *Voice in Qualitative Inquiry: Challenging Conventional, Interpretive, and Critical Conceptions in Qualitative Research* (with Lisa A. Mazzei; Routledge, 2009). Her work has also been published in *The International Journal of Qualitative Studies in Education*, *Qualitative Inquiry*, *The International Review of Qualitative Research*, and *Qualitative Research*.

Aaron M. Kuntz is Professor and Department Head of Educational Studies at the University of Alabama, where he teaches graduate courses in qualitative inquiry and foundations of education. His work has appeared in such diverse journals as *Qualitative Inquiry*, *Cultural Studies ↔ Critical Methodologies*, *The Journal of Higher Education*, *The Review of Higher Education*, *International Journal of Qualitative Studies in Education*, *Educational Studies*, and others, as well as in numerous book chapters. Alongside colleagues from the Disruptive Dialogue Project, he co-authored *Qualitative Inquiry for Equity in Higher Education: Methodological Implications, Negotiations, and Responsibilities* (Jossey-Bass Publishers). Dr. Kuntz's co-edited volume *Citizenship Education: Global Perspectives, Local Practices* (with John Petrovic) was published in 2014 with Routledge Press. His most recent book, *The Responsible Methodologist: Inquiry, Truth-Telling, and Social Justice* (Routledge) was selected as Honorable Mention for the 2017 AERA Qual SIG book award.

Lisa A. Mazzei is Associate Professor in the Department of Education Studies at the University of Oregon. She is the author most recently of *Thinking with Theory in Qualitative Research: Viewing Data Across Multiple Perspectives* (with Alecia Y. Jackson; Routledge, 2012), *Voice in Qualitative Inquiry: Challenging Conventional, Interpretive, and Critical Conceptions in Qualitative Research* (with Alecia Y. Jackson; Routledge, 2009), and *Inhabited Silence in Qualitative Research: Putting Poststructural Theory to Work* (Peter Lang, 2007). Her work has also appeared in leading journals such as *Educational Researcher*, *British Education Research Journal*, *The International Journal of Qualitative Studies in Education*, and *Qualitative Inquiry*.

Wilson Okello is a Visiting Assistant Professor of Black World Studies in the Department of Global and Intercultural Studies at Miami (OH) University and Founder of the Truth to Power Project. Bridging the scholar-artist divide, his research employs Black feminism(s) and aesthetics to critique and advance meaning-making theory and pedagogical praxis. His work has been published in venues such as *Journal of College Student Development* and *Journal of Curriculum and Pedagogy*.

Tami Spry is Professor of Performance Studies in the Communication Studies Department at St. Cloud State University in Minnesota. She employs auto-ethnographic writing and performance as a critical method of inquiry into diffi-cult sociocultural issues; specifically, her performance work and publications engage issues of race, sexual assault, grief, shamanism, and mental illness. She performs in the United States and abroad, including recently at the University of Bristol, the University of Cambridge, and the University of Oxford. She is the author most recently of: *Body, Paper, Stage: Writing and performing autoethnog-raphy* (Routledge, 2011) and *Autoethnography and the Other: Unsettling power through utopian performatives* (Routledge, 2016), and *How Writing Touches: An intimate collaboration* (Cambridge, 2012; with Ken Gale, Ron Pelias, Larry Russell, and Jonathan Wyatt).

Karen M. Staller is Associate Professor of Social Work at the University of Michigan. She received her educational training at Cornell Law School and Columbia University School of Social Work, where her dissertation on runa-way and homeless youth was awarded with distinction. She practiced public interest law with low-income senior citizens and at-risk adolescents in New York City. Her scholarship focuses primarily on runaway and homeless youth (and other at-risk adolescents). She is interested in the complicated interplay between social problem construction, social service delivery, and social policy. She is the author of *Runaway: How the Sixties counterculture shaped today's practices and politics* (2006, Columbia University Press), co-editor (with K. Faller) of *Seeking Justice in Child Sexual Abuse: Shifting burdens and sharing responsibilities* (2009, Columbia University Press). She is currently working on a book titled *Extra! Extra! Read All about It: The newsboys housing lodge and Mr. Brace's Chil-dren's Aid Society* (forthcoming, Oxford University Press).

INDEX